CHANGE-READY OR NOT

**Climate:
Our Choice,
Our Responsibility**

First published in Australia 2015
Revised edition: August 2018
By Alan Garman

Cover Photo: Alan Garman

Cover design and layout: Dan Hickingbotham – www.alined.com
Printed by Lightning Source

ISBN 978-0-9944144-0-3

*To my family and close friends whose encouragement and support
contributed so much to this book.*

*I dedicate this work to all of us who call this
beautiful and bountiful planet home.*

August 2018 Update

*Change — Ready or Not scopes possible solutions to our continuing
alteration of the chemistry of the atmosphere and oceans and the enduring
contest between our consumer society and the environment.*

(From Author's Note: Page 229)

Contents

Foreword

Tim Costello, CEO – World Vision Australia

With this book Alan Garman has made a valuable contribution to the national and global conversation about the climate and what we need to do about it.

Alan has adopted a thinking framework that combines optimism with urgency, and a lucid exposition of the science of climate change with exploration of the responses that are possible and necessary. In this he makes use of the insights that come from engineering, management, political and even spiritual disciplines and perspectives.

This is an account of where we stand on climate and what needs to happen next. While informed by science, it is not a scientific work or even an academic one. It is a personal view – rational, logical, persuasive, but one that is consciously directed at shifting the view of those in positions of power and influence who are yet to comprehend the vastness of the challenge that confronts humanity.

While oceans of ink have been spilt in climate debates, there is still a space and indeed a necessity to make this a democratic conversation. Science illuminates the truth, but the action required if we are to avoid catastrophic consequences lie very much in the political and economic spaces, where all of us have a stake and a role to play.

I'm thankful that with this book Alan has made such a strong and insightful intervention. I am sure many will read it with interest and a growing concern and determination to help turn the tide of indifference and inaction. But the window of opportunity is short, so as Alan concludes, the time to start is now.

September 2015

Preface

As a boy, I sometimes found myself reflecting on how powerful civilizations, such as those that existed in Mesopotamia, Egypt and Rome had failed. It was beyond a young boy's comprehension that a similar fate could ever befall us. Was this youthful naivety? The rise and fall of civilizations is not necessarily confined to the ancient world and the risks faced will be different in each case.

We are now, as global citizens, in a situation where the danger being faced is the very environment that sustains us. It is questionable whether the highly complex and potentially fragile social order that is our world society, could withstand the stresses of a genuinely hostile planetary environment.

We are all bound to this Earth and to each other. We have enjoyed the bounty and the joys of this planet for millennia and if the time has come to put 'shoulders to the wheel', those shoulders will be ours.

This book will focus on strategies and changes to the prevailing economic paradigm that will be essential to combat global warming. Changes I believe the current political and economic debate is failing to address. Not all details of climate science and scientific research that are already available in the public domain will be addressed, as this is a conversation that explores many other aspects of the climate debate. Social, political, economic, scientific and engineering issues are all discussed and supported as required.

The co-dependent problems of rapidly increasing greenhouse[1] gas emissions, changes to the chemical composition of the atmosphere and oceans plus gradually increasing global temperatures are multi-faceted.

One contributor to the increase in emissions worthy of special mention is the continually expanding world

---

[1] The 'greenhouse' effect is described in Chapter 2 - Habitat.

population, which some argue to be a root cause of global warming and climate change. While there is truth in the argument that the larger the population, the greater the emissions, it is an oversimplification to suggest that controlling population alone will provide a timely solution to global warming. Even strategies to stabilise (let alone reduce) the world population would encounter deep complexities such as poverty and differing cultural, religious, political and social justice systems, conventions and traditions.

There is also the matter of education. It is commonly accepted that there is an inverse correlation between higher levels of education and birth rate. In countries rich and poor, it has been widely observed that a higher level of education, particularly of girls and young women, has a multiplier effect on the well-being of society. There is also a corresponding reduction in the birth rate, which is usually beneficial for the environment.

Adding to the complexity is that higher levels of education generally lead to an understandable desire for a better standard of living, resulting in increased consumer demand which *under current economic paradigms*, will lead to even higher emissions and worsening global warming.

I take the view that at any point in time, the population of Earth is a given that we must accept. We all have an equal right to be here and the solution to global warming must accommodate that reality. Development of 'zero net emission' renewable energy and fuel systems will be exceedingly challenging, but will ultimately prove more practical than hoping to significantly reduce the Earth's population in a time-frame meaningful enough to counter the currently observed chemical changes in atmosphere and oceans. Similarly, the benefits of a better standard of living are a reasonable aim for all people, so we must work towards economic paradigms that can deliver this and also underpin a stable environment.

It is therefore my contention that we must look to economics, politics, science and technology to provide the tools to stabilise world temperature, rather than blame it on world population levels. I am not of course, suggesting the question of population should not be addressed and preferably in parallel with fitting economic and technological climate solutions, but strategies for stabilising and ultimately reducing the world population are beyond the scope of this book.

Finally, together with examples from around the world, this book includes a number that are specific to Australia. While these reflect situations familiar to me, it is also relevant that there are powerful and influential positions being articulated in Australia that are a negative for effective action on global warming.

These negative perspectives are not confined to any one country, so the Australian experience is representative of opinions that may be heard in other parts of the world. If this were not the case, there would be no requirement for further commentary on global warming or climate change and we would by now be witnessing massive world-wide greenhouse gas mitigation measures—something that is clearly not happening.

At the time of my boyhood reflections, atmospheric carbon dioxide (CO_2) had only increased by about 10% from the dawn of the Industrial Era, not the massive 43% we see now. This book is an expression of opinion based on a lifetime of education, commenced in 1944 and continued through the decades, with the most recent qualification completed in 2012. Studies in physics, applied mechanics, thermodynamics, mathematics and the humanities have been supplemented by observations, readings and discussions with like-minded and contra-minded people.

On a daily basis, we absorb news articles and reports around climate and climate change. Many of us watch the evening news to update ourselves on weather events of the day and the near future, but few delve into the specifics of long-term climate prediction and attribution of cause.

Assertions made in this book are a result of many years of study, have veracity and I believe, are verifiable. Strategies, plans and projections for a future that is yet to unfold, are also included and can only be confirmed or otherwise, by the passage of time. New ideas are not always able to be easily tested. You the reader will become expert at classifying those sections of this work that crystallise immediately with your own worldview, from those sections that will require further analysis and discussion.

This work is not a substitute for the thousands of brilliant scientific reports, books and journals written about global warming and climate change which define the situation. It is about what human beings could be doing to address these matters.

My view is that the vast and overwhelming weight of scientific research and analysis pertaining to the physics of the atmosphere is largely agreed upon. However, not everyone concurs with the view that CO_2 emissions emanating from human industrial and agricultural activity will create a potentially devastating problem for humanity. Helping to change the dominant discourse on the emerging crisis in the atmosphere/biosphere is the powerful motivation for this book.

To change the perspective of those who continue to deny human involvement in the powerful structural changes taking place in the chemistry of the atmosphere/biosphere will be difficult. But at the heart of it, effective action on global warming will ultimately depend on those who manage—or influence those who manage—the politics and economics of our society.

That is us—and that is a job for us all.

Introduction

We are a privileged people, living in a fortunate age and we are at an historic crossroad. One road looks beguilingly easy to travel. This is the path to ever-increasing greenhouse gas emissions, continuing on until eventually, we simply run out of road. The other road, leading to a net zero carbon dioxide emission economy, is unmapped and harder to gauge. Our moral dilemma is whether we continue to take the easy road and live as if there is no tomorrow, a contemporary version of 'fiddling while Rome burns', or whether we recognise that our challenge and opportunity of the present day is to explore the unmapped path and make possible a sustainable future and continue this fortunate age.

Our current management of climate change appears to make the unexamined assumption that the world's decision-makers know the exact limit to which the biosphere[2] of the Earth can be pushed before something catastrophic happens. There is no science to support the position that we can continue to increase atmospheric carbon dioxide (CO_2), without some negative effect on long-term climate stability. We cannot assume that no catastrophic event will occur as a result of our inappropriate management of the environment. In fact, current science warns us that there is a limit. The crossroad we face is another opportunity to call on human ingenuity to achieve something great. We know this can be done; as it has been, many times throughout human history.

If we don't act upon the known science and its predictions and continue instead on our current path, we are all taking a huge gamble by behaving in a manner that seems to assume that nothing bad will ever happen.

---

[2] The part of the planet supporting life; includes air, land, and water. The biosphere is a total of all ecosystems on the planet.

Civilization is a delicately balanced circumstance. In a system as large and complex as the biosphere of the Earth, the risk is that variations in the chemistry of the atmosphere will be magnified by unpredictable, non-linear feedback loops, which could in time have devastating consequences for the finely-balanced organisational structure of human society.

In 1765[3,] all atmospheric greenhouse gases amounted to only about 280 ppm[4], which equated to 0.028% of the dry atmosphere. Today these gases have increased to just over 400 ppm or approximately 0.04% of the dry atmosphere—a whopping 43% increase. Another way of considering this is that in 1765, the Earth was maintained at just the right temperature by 28 thousandths of 1% of the atmosphere being composed of greenhouse gases. The 43% greenhouse gas increase in the last 250 years has seen this amount increase to just over 40 thousandths of 1%.

Climate change sceptics sometimes claim the increase over the Industrial Era is only 0.01% of the overall atmosphere and is therefore insignificant, but they are wrong on two counts. The first is that the change is not 0.01%, but 0.012%[5], which is a 20% rounding—on the 'wrong' side of prudence. The second and more important reason they are in error is that the overall quantum of greenhouse gases in the atmosphere (that keep the planet from being either a frozen 'ice-ball' or dangerously warm for the proper functioning of our civil society) was so very tiny in the first place; namely, less than three one hundredths[6] of one percent of the entire atmosphere.

Although the change in greenhouse gases is numerically small, it is a very large percentage change. The astonishing reality is that such a tiny measure of greenhouse gas has kept

---

[3] James Watt's invention of the modern steam engine in 1765 is taken as a practical starting date for the modern industrial era.

[4] Parts per million is abbreviated as ppm.

[5] 400 ppm minus 280 ppm = 120 ppm, therefore = 120/1,000,000*100 = 0.012%.

[6] Previously referenced: in 1765 greenhouse gases were 280 ppm = 0.028% = less than 0.03%.

the planet at just the right temperature for the whole of human history and for hundreds of thousands of years before human history began. The planetary 'sweet spot' is frighteningly small.

As one of the seven billion of Earth's human stakeholders, I believe that given the risks, we must insure our future by *ensuring* that policy responses and outcomes are proportionate to the disruption that would be caused by worst-case scenarios within the biosphere.

To meet the challenge of providing sustainability for all citizens, now and in the future, immediate action must be taken on **nine critical climate protection measures:**

1. *Develop a sense of urgency* within the community, to build consensus and implement policies designed to stabilise the chemistry of the atmosphere

2. *Reframe the dominant public discussion.* Stimulate debate on ways to limit technologies that are reliant on fossil fuel combustion

3. *Expedite development of future technologies.* There will be a profusion of zero emission fuel, energy, transport and manufacturing capability by the end of this century, but this may be too late. We need to prioritise the establishment of a goal to implement the non-carbon energy systems of the latter half of this century to 2030 or earlier

4. *Fund development of future technologies* now to a level that matches the gravity of the emerging climate crisis and the complexity, magnitude and urgency of the task

5. *Harness the power of technological compounding.* Developments in science and technology mostly compound on a previous level of capability. Advances in the efficient use of fossil fuels have been compounding for two centuries and have led to many remarkable applications. The challenge is now to achieve similar outcomes for non-fossil fuel/renewable energy applications in a greatly compressed timeframe

6. *Inclusion of CO_2 emission costs.* Ensure all goods produced and services delivered anywhere in the world incorporate a CO_2 emission cost. Employ high-level computing capacity to achieve this and use the data to implement a globally consistent model for pricing emissions
7. *Employ an economic gradualist approach.* Counter-intuitive though this may seem, incremental change may be necessary to make palatable the innovative solutions to the economic challenges, which may otherwise seem overwhelming, as we attempt to fund effective climate change mitigation measures
8. *Manage the issue of global warming using hazard analysis and risk assessment tools* in a manner consistent with the possibility and consequences of planet-wide environmental catastrophe
9. *Inspire worldwide support, at all levels of society, for the changes* that are critical for our continued security and enjoyment of family, business and leisure.

The achievement of chemical stability within the atmosphere and the oceans are the two critical geophysical factors on which current generations will be judged, twenty years from now.

This task is substantial. As a people, we perhaps have no more than five years to make some very important decisions. The political and economic decisions that are made within these next few years will settle the debate. The complex issues raised in this book could either trigger a sense of alarm, or engender a sense of determination to meet the challenges posed by climate change.

My position is above all, one of optimism. I believe with political determination and substantial global action, we do have every chance of arresting the ever-increasing concentration of atmospheric greenhouse gases.

1. Consilience

Hundreds of independent lines of research, using entirely autonomous data sets, unite in what might, at some future date, be called 'The Anthropogenic[7] Climate Theory'—a beautiful consilience[8] of scientific evidence. Many of the theories we accept today are not proven by a single scientific fact, but by a convergence of many.

Darwin's *Theory of Evolution* and the *Theory of Continental Drift and Plate-Tectonics*, are two such examples. We should now be elevating the converging and convincing research on the role of human-induced changes to the Earth's atmosphere and oceans, to the status of these two theories.

But instead, we are still divided on the issue of whether the planet is warming and the likely causes. I believe this is because any opinion that suggests we are not responsible for atmospheric and oceanic changes are promulgated with such vigour that people can mistake such views as scientific disagreement on global warming. This is not true. What we are seeing and hearing are simply the same few scientifically incorrect statements, being presented as facts and repeated over and over again.

This is definitely not a consilience of scientific evidence.

So do we need more text on climate change? Hasn't enough been said already? Surely we are saturated with information? If we use the yardstick of how many truly effective greenhouse gas reduction strategies are currently deployed, it is very clear that we must do significantly more to reverse the climatic trends now emerging. Until approximately 1990, the increase of carbon in the atmosphere, attributable to

---

[7] Anthropogenic: pertaining to phenomena that are a direct, or indirect, result of human activity.
[8] Consilience: independent lines of scientific research/evidence that converge to produce a unified theory.

11

human influence, was largely inadvertent. Today, it is part of a de facto, but in some ways almost deliberate, economic policy.

We have known the danger of exponentially increasing greenhouse gases for more than twenty five years and yet there is still no effective global action. Identifying the problem is insufficient; real change will depend on successfully formulating and implementing solutions and this cannot occur without effectively communicating the need for such action.

It is worth noting that those who oppose substantial action to mitigate global warming certainly understand the critical importance of broadcasting their views. They successfully promote the conflicting notions that global warming is either not real or if real, then nothing to worry about. This is extended to claim that nothing needs, or can be done to reverse the warming trend because it is simply part of a natural cycle.

We know that planetary cycles do occur, but to conclude that the 43% increase in the powerful greenhouse gas, carbon dioxide (CO_2), during the Industrial Era is a naturally occurring phenomenon is incomprehensible and cannot be the basis of any viable political or economic strategy.

Removing carbon from the atmosphere and de-acidifying the oceans will be a massive endeavour, but human society has faced and beaten immense challenges in the past. During the Second World War, the warring nations combined spent up to 40% of Gross World Product (GWP)[9] to the war effort. The required expenditure was found and global economic structures survived and ultimately prospered. If we adopt this approach and prevent further damage to the atmosphere and oceans, then the world's economic and social structures will continue to survive and prosper.

There are many precedents of science and engineering being expedited by adequate funding of very tight, seemingly

---

[9] Gross World Product (GWP) is the combined Gross National Product of every country in the world.

impossible, deadlines. The Manhattan and Apollo projects are two of the many.

With appropriate economic structures in place it is not unrealistic to believe that we can remove fossil fuels (and carbon) from our energy and fuel cycles in a time-frame that is rapid enough to preserve our environment in a condition remaining benign to our civilization.

Eliminating CO_2 emissions from our industrial processes will not occur through some form of community osmosis. Status quo politics has no place in the twenty-first century. All people who recognise the dangers posed by human-induced global warming must learn from the marketing approach of those who deny a human role in that process and become equally vocal in delivering their own message. To effectively counter the negative argument, we need many more people reinforcing the peer-reviewed scientific truth that the planet is indeed warming.

There is a curious anomaly in matters related to climate change. Many members of the public and an influential number of those who manage the world's social and economic affairs are dismissive of the message from scientists who work in fields relevant to the atmospheric and oceanic changes we observe. They appear not to trust what the science is saying.

This curiosity is heightened when we ponder the fact that in a passenger jet, 12,000 metres above the Earth, we all trust our very lives to the scientific research and skill of the scientists and engineers that made the aircraft possible in the first place. When we are sick, we place our trust in the scientific research behind the medical interventions that heal us and that we call upon to save our lives.

We should now adopt a similar position in relation to global warming and climate change. When the political and economic issues are finally settled, it is and will always be a scientific and engineering problem.

2. Habitat

We are destroying our habitat.

The primary focus of the climate change debate should not only be about the habitat of the world's coral reefs, or the polar bear. It must be about the habitat of the human being. Of course, it is to our great peril if we disregard the role of the natural world, both from the obvious ethical perspective and as a canary in the mine[10] for our manufactured world.

The survival of polar bears, coral reefs and the rest of the natural world are threatened by the same forces that threaten humans. It is therefore essential to maintain the habitat of humans, in a condition as close as possible to that existing prior to the Industrial Era, not only for the benefit of ourselves, but for all other living creatures.

Easter Island

While it is true that we may never be sure of the exact cause, there is still a lesson for contemporary humanity in the events that led to the depopulation of Easter Island some four centuries ago. There are several explanations offered for the demise of this previously flourishing population.

The theories range from overpopulation, a tsunami or smallpox epidemic, rat plagues and/or deforestation through excessive felling of the native palm trees either to produce rollers to facilitate the quarrying, carving and transporting of the monolithic stone figures for which Easter Island is famous, or simply to obtain the edible palm hearts.

It is possible that a combination of these factors led to the depopulation of Easter Island. Tsunamis aside, massive felling of fruit-bearing adult palms for whatever reason, rats gnawing

---

[10] In the early days of coal mining, miners would often take a caged canary underground with them to give an early warning of the presence of dangerous methane gas.

at a decreasing number of palm seed shells, disease and over-use of resources are all a result of human interaction with the natural world. This has everything to do with 'tipping points'[11]. At some stage a tipping point would have been passed and significant stands of fully grown, fruit-bearing palms critically diminished and the structure of the Island's society would have broken down.

It is possible to foresee that far, far into the future, an archaeologist examining the evidence of the long-vanished great societies of today, may well pose a question such as: 'What were 21st century people thinking, when they allowed the level of carbon dioxide in the atmosphere to increase to that terminal level?' Of course, the answer is that there will never be a last critical increase. Before that last tiny hypothetical increment in parts per million of carbon dioxide occurs, the tipping point will have passed, making the atmosphere and oceans progressively less able to maintain our complex civilisation, long nurtured by a benign climate system.

While people may continue to debate the final fate of the early inhabitants of Easter Island, there is a prescience and relevance for our society today. The case against over exploitation of resources stands.

Fast Forward Four Centuries

If this seems too disconnected from our everyday lives, then perhaps something more tangible and alarming is the approximate 30% increase in acidification of the world's oceans since the start of the Industrial Era. This threatens the entire marine environment and is a direct result of the oceans absorbing a portion of the extra carbon dioxide (CO_2) that is being pumped into the atmosphere. Issues of ocean

---

[11] A 'tipping point' is the point at which some entity is displaced from a state of stable equilibrium into a new and different state.

acidification are discussed further in Chapter 10: Science and Engineering – Models and other Information.

A Wasting Resource

A significant problem of the early 21st century is that we live in a right now society; a society where people expect things to happen almost instantly. The fact is, right now, in 2015, generally things appear to be fine. Life goes on. People are busy with day-to-day issues; too busy to worry about global warming. Sure, some people are aware of aspects of the climate that are not quite as they were fifty years ago.

There do seem to be more extreme weather events happening, but overall they haven't yet had a significant impact on our day-to-day lives. In any case, the whole problem seems too big and too difficult. How would we ever start sorting out such complexity?

To the casual observer living in a First World country, there appears to be no immediate threat arising from inadvertently altering the chemical composition of the atmosphere. In the shopping malls, cinemas and in our automobiles, on jet skis and on holiday, very little changes— certainly nothing that would indicate an impending crisis of habitat. If anything, it has all become more affordable, at least to those fortunate enough to live in First World countries and who are either employed in secure, well-paid jobs or have independent means.

Another factor that works against real and effective action to reduce CO_2 emissions is the political influence of—and resistance from—large commercial and industrial interests. With both consumers and business comfortable with the status quo, there is little political incentive for global action to reduce emissions commensurate with the scale of the problem, or the magnitude of the risk.

The sense of no immediate risk and the presence of widespread resistance, allow environmental issues to be easily hijacked into a debate on side issues such as how much tax we should be paying. It is astonishing that while we accept the necessity to pay a fee for the privilege of dumping household rubbish into landfill, loud protests erupt at any proposal for a fee to dump CO_2 into the atmosphere.

The main difficulty in obtaining substantial action on climate change appears to be that it doesn't have an immediate and observable impact on our day-to-day lives. But, what if we knew the exact tipping point, beyond which the climate would become uncontrollably hostile to society?

There is a collective view that 450 ppm atmospheric CO_2 is the maximum concentration if we are to limit the average global temperature rise to 2° Celsius (2°C)[12]. This figure has been widely discussed in the public arena and will be discussed further in this book. However, what if 450 ppm atmospheric CO_2 was *actually the tipping point* for an uncontrollably hostile climate? Surely, with the knowledge that the level of CO_2 in the atmosphere has already increased from pre-industrial 280 ppm to over 400 ppm, we would now be seeing a far greater sense of urgency in remedying the situation.

An impediment to achieving the action necessary on climate change is that people, business and governments are taking positions that are not based on climate science, chemistry and physics. Global warming and climate change was seen as a scientific problem 25 years ago, but in the public consciousness, it is now more of a social, cultural, economic and political issue. Many people now take positions regarding the chemistry of the atmosphere and oceans that are influenced by political, religious, cultural or socio-economic grouping and the perceived economic cost to a particular group. The work of climate scientists, based on scientific observations and data, is effectively dismissed.

There is also a continuing belief that no real change is necessary to our current methods of using resources. It is

---

[12] 2° Celsius equals 3.6° Fahrenheit (3.6°F)

naive to believe that the continued use of processes that pay scant regard to the sustainability of those resources remain appropriate. In the past, this resource management strategy appeared to work reasonably well, only because the effects of this strategy were not clearly understood, identified or measured.

Today we have much greater understanding. News reports regularly inform us that scientific predictions for warming trends and polar ice loss are being met and more alarmingly, exceeded. There is no longer any excuse for not recognising that the most important wasting resource is now the ability of the atmosphere and the ocean to act as an unlimited sink[13] for CO_2 emissions.

Unfortunately, the capacity of the atmosphere and the ocean to act as a CO_2 emission sink is limited; assuming that we wish to maintain stability in both the biosphere and our habitat.

In Britain, from September 1939 to May 1940, there was a period known as the 'Phoney War'[14]. The Phoney War ended abruptly when the bombs started dropping. What is happening now, relative to global warming and climate change has all the hallmarks of a 'Phoney Climate War'. The 'climate version' of the 'Phoney War' will no doubt also end, as the real problems become undeniable. The sooner we recognise global warming for what it is—a direct threat to our current way of life—the more chance we have to avoid unfavourable outcomes.

---

[13] A carbon sink is a system/repository that will absorb and store carbon dioxide/carbon, such as a forest, an ocean or the atmosphere.

[14] In those early months of the Second World War, there was no observable direct aggressive action against people living in British towns and in the countryside. It was certainly not a 'Phoney' war for people in Poland and other parts of continental Europe.

The Greenhouse Effect

A certain level of *'greenhouse effect'* is essential to support our civilization. The greenhouse effect is a natural phenomenon necessary to sustain human life on this planet. Certain gases in the atmosphere (the greenhouse gases) act in exactly the same manner as the common backyard greenhouse by preventing some of the Sun's heat being radiated back into space. It is estimated that the average world temperature would be 33°C cooler if not for the greenhouse effect. This would mean that world average temperature would be around minus 19°C, rather than the current (plus) 14°C.

Calculated to four decimal places, nitrogen, oxygen and argon compose 99.9578% of the dry atmosphere by volume. These are not greenhouse gases. The remaining 0.0422% is made up of greenhouse and other trace gases. Carbon dioxide (CO_2) is the most prolific of these greenhouse gases and currently accounts for over 0.04%[15] of the dry atmosphere. The other dry atmosphere greenhouse gases include methane, nitrous oxide, ozone and the halocarbons. In addition to the dry atmosphere greenhouse gases, water vapour is another significant greenhouse gas that will be discussed later in this section.

The global warming potential of a gas is determined by the greenhouse potency of the gas and the time it remains in the atmosphere. The resulting warming effect of any gas—its global warming potential—is calculated from its greenhouse potency and the length of time it remains in the atmosphere.

CO_2 is not the most potent of the greenhouse gases, but it is by far the most prolific. It also remains in the atmosphere for up to 200 years. The increasing levels of CO_2 emanating from human industrial and agricultural processes make it the largest contributor to global warming.

It is also important to remember that emissions from human activity are constantly adding to the naturally occurring

---

[15] 400 ppm = 0.04%.

emissions, thus increasing the total pool of greenhouse gases. The increase in atmospheric CO_2, over the industrial era, correlates more-or-less exactly with the increased human industrial and agricultural activity over the same period.

Carbon Dioxide Equivalent

The global warming potential of greenhouse gases such as methane, nitrous oxide and the halocarbons[16] vary markedly from CO_2. They vary in both potency and the length of time they remain in the atmosphere. Compared to the quantity of CO_2 in the atmosphere, the quantity of the other greenhouse gases is small to insignificant. Nevertheless, their global warming potential is still significant due to their greater potency. To represent this in a single measure, the normal convention is to specify the amount of CO_2 that would be required to produce an equivalent warming effect.

A measure called the 'carbon dioxide equivalent' (CO_{2e}) is a method by which the total global warming potential of all the greenhouse gases[17] can be evaluated. While there appears to be broad consensus on the actual CO_2 in the atmosphere in parts per million, obtaining a definitive value for CO_2e can be more problematic than for CO_2 alone.

Adding to the complexity, some industrial processes produce various particulates, aerosols and sulphur dioxide (SO_2) and these can have a cooling effect by reflecting more of the Sun's radiation back into space. However, the cooling effects of aerosols, particulates and SO_2 do not significantly offset the warming effect of the major greenhouse gas—CO_2.

The sole mechanism, by which the Earth is either warmed or cooled, is radiated energy. The Earth is warmed by solar

---

[16] The principal halocarbons are CFCs (chlorofluorocarbons), HCFCs (hydro chlorofluorocarbons) and HFCs (hydro fluorocarbons) respectively.

[17] CO_2e includes the warming potential of carbon dioxide as well as methane, nitrous oxide and the halocarbons. These gases are often referred to as 'well-mixed' greenhouse gases because they are homogenous worldwide.

radiation from the Sun. Our comfortable global temperature is maintained because a proportion of incoming solar radiation is reflected back into space by the greenhouse effect, the particular chemistry/composition of the upper atmosphere or by particulates, aerosols and sulphur dioxide —or it is radiated back into space as infra-red radiation, from the Earth's surface. There is a delicate balance between incoming and outgoing radiated energy and the resulting net warming effect on the planet is measured in watts per square metre of the Earth's surface.

So, the positive or 'warming' component of the equation is net amount of all incoming solar radiation that is 'retained' in the environment. All naturally occurring and anthropogenic influences on the Earth's 'greenhouse' are included in the calculation of net warming effect. The added CO_2 and other greenhouse gases from industrial and agricultural origin, amplify the greenhouse effect.

The negative or 'cooling' component of the equation is made up of the amount of solar radiation that is radiated back into space by the mechanisms described above and, as a consequence, is 'not retained' in the environment. The directly radiated infra-red component from the Earth's surface depends on the average surface temperature of the planet measured on an absolute scale of temperature[18]. The reflected component is amplified by atmospheric particulates, aerosols and sulphur dioxide, of both naturally occurring and anthropogenic origin.

The difficulty in obtaining a definitive value for CO_2e appears to stem from two interpretations of exactly what should be included in the measurement. One interpretation of CO_2e is taken as the summation of only the warming effect of *all* greenhouse gases in the atmosphere, expressed in terms of CO_2. In this interpretation, the effects of the various particulates, aerosols and sulphur dioxide are excluded.

---

[18] The absolute scale of temperature in the SI system of units is the Kelvin.

The other interpretation of CO_2e includes the effects of atmospheric aerosols and particulates in addition to all greenhouse gases[19]. While this method appears to be more complete, it does raise some issues which are difficult to resolve. These issues result from the considerable variation in the estimated cooling effect of atmospheric aerosols, leading to consequential variations in estimates for an all-inclusive and precise global warming potential.

Aerosols and other things in the Air

The term aerosol defines a large range of particulate matter from very small particles of solids such as black carbon (soot) and dust, sulphates or liquids, which can vary in size from 1/100th of a micrometre[20] (μm) to 100 micrometre (μm). Aerosols originating from anthropogenic sources usually remain in the atmosphere for only a few days to a few weeks, meaning that uniform mixing on a global scale is not possible.

Aerosols are mostly concentrated in the Northern Hemisphere and are associated with extensive centres of industrial production, such as Northeast Asia, Eastern United States and Europe. In contrast, the 'well-mixed'[21] greenhouse gases (GHGs—carbon dioxide, methane, nitrous oxide and the halocarbons) are distributed throughout the globe. The fact that aerosol pollution is short-lived and more prevalent in the Northern Hemisphere makes it impossible to obtain a well-mixed and accurate global average for particulates and aerosols.

Additionally, when black carbon/soot is deposited in the Polar Regions (on ice shelves, sea ice, snow or snow-covered

---

[19] Whether naturally occurring, or of anthropogenic origin.

[20] One micrometre is equal to one micron. The term micron is still widely used, but not in the SI system of units. In the SI system: one micrometre = one millionth of a metre = one thousandth of a millimetre.

[21] Gases such as carbon dioxide, methane and nitrous oxide stay in the atmosphere for long enough (up to 200 years for CO_2) to become homogenous around the world – hence the term 'well-mixed'.

ice) it will reduce the reflectivity—albedo[22]. This allows more solar radiation to be absorbed, which will increase global warming and the rate of ice-melt.

In summary, the calculation of an exact value for aerosol pollution is problematic because both output[23] and the atmospheric concentration can vary from day to day. Of course, the same applies to CO_2, but daily variations are far more significant in particles that may only remain in the atmosphere for a few days or a few weeks, than daily fluctuations in a gas (CO_2) that will remain in the atmosphere for up to 200 years.

The vastly different time-span that aerosols remain in the atmosphere, relative to CO_2, poses an additional ironic dilemma. In time, aerosol emissions from resource extraction, industrial production and energy generation will be reduced, as part of much needed pollution reduction[24], for health and other environmental considerations. The cooling component of aerosol and particulate pollution, that may currently be partially offsetting the warming effect of the well-mixed GHGs, will be correspondingly reduced. However, the well-mixed GHGs, carbon dioxide, methane, nitrous oxide and the halocarbons, will remain, effectively increasing the rate of warming. This is another strong incentive for humanity to avoid continually adding to the CO_2 molecules in the atmosphere.

Hence, carbon dioxide, methane, the nitrous oxides, the halocarbons and a few of the aerosols and particulates increase average global temperatures. At the same time, aerosols and particulates mostly have a cooling effect. The

---

[22] The Albedo Effect (the proportion of solar radiation reflected back into space) is explained in Chapter 8: Melting Ice.

[23] The principle sources are: Industrial Production, the Mining and Extractive Industries, Land Transport (Diesel, Gasoline and LPG etc.), Shipping and Aviation.

[24] See Chapter 10; Science and Engineering – Models and other Information, for more on the 'clean-up' of aerosols and particulates.

summation of all these different inputs influences the previously discussed *carbon dioxide equivalent (CO₂e)*.

Risky Experiment

My assessment is that the warming factors massively outweigh the relatively insignificant cooling factors. Furthermore, when all anthropogenic warming and cooling factors are considered, the CO_2e is likely to be at least equal to the widely canvassed 450 ppm for CO_2. It is my view that 450 ppm CO_2e alone is significant and a corresponding global temperature rise of at least 2°C is already a very likely outcome.

We have embarked on a very dangerous global environmental experiment.

I have noticed recent news items that reference certain dates in connection with CO_2 levels that should not be exceeded, if our aim is to keep global temperature rise to less than 2°C. There are various dates given by which these targets should be met, with 2050 and 2100 often mentioned.

I think there is great danger in even mentioning a date such as 2100. In the context of the current challenges that global warming is presenting, action is required now—not in the future. Using James Watt's invention of the modern steam engine in 1765 as a base for our industrial age, 2100 represents a 335 year time-span for action. My concern is that the global community's focus may be distracted from the reduction in global emissions that must be achieved before, for example 2020/25, to targets much further into the future; targets that are so much easier to ignore in the short term.

The additional danger in even talking about dates so far into the future is that global average CO_2 levels are increasing exponentially. It has taken about 250 years for atmospheric CO_2 to increase to current levels from the pre-industrial level. My reading of the situation is that the first half of this increase took well over 200 years and the next half took only about 40 years.

It is open to question as to whether the responsibility for meeting date and emission reduction targets rest with the global community at large, world leaders, or with the business, commercial and industrial interests that have a major input into the way in which the global economy is managed.

In summary, in reference to CO_2 levels, it must be remembered we are dealing with the environment of a planet which has evolved over millions of years and that mankind has altered that environment over only about 250 years —maximum!

Whatever time-span is being considered, the underlying thesis is unaltered. The Earth is warming, the Polar Ice is melting and the climate is changing all over the world.

CO_2 or CO_2 Equivalent

Notwithstanding the potentially great value of CO_2e in understanding the combined warming potential of all atmospheric greenhouse gases and other agents, this book will mostly adhere to the common practice of referring only to CO_2. The rationale for this decision is that while a CO_2e value can be calculated for the warming effect of the well-mixed GHGs (carbon dioxide, methane, nitrous oxide and the halocarbons) the same is not true for the previously discussed particulates and aerosols.

This means there is a general lack of agreement, on a definitive value for CO_2e, which includes both the positive and negative effects of particulates and aerosols. Also, CO_2 is the parameter most frequently used in the wider community and, of course, CO_2 is the principal greenhouse gas. However, any reference to urgently reducing CO_2 emissions, made throughout this book, applies to other greenhouse gases as well.

Highest CO_2 since 'Far-Too-Long-Ago' for Comfort

From my reading of the situation, I accept that the current CO_2 level of just over 400 ppm has not been seen on Earth for at least 800,000 years. In truth it is more likely than not, that our current level of CO_2 is higher than has been seen for two, five or even more millions of years.

There is no comfort at all in this knowledge. We are told that in those times global temperatures and the planetary environment were vastly different to those that support our complex society today. Our own common sense tells us not to test such extremes of CO_2 and the environment.

The increase in CO_2, from the pre-industrial level of 280 ppm to the current level of just over 400 ppm, is extraordinary and is certainly not part of any kind of 'natural' balance in the biosphere. The increase of 43% over the Industrial Age is alarming. The increase of approximately 33%[25] since the start of the 20th Century is even more alarming.

Many climate commentators and scientists have stated that to avoid dangerously high temperatures we should not tolerate a CO_2 level greater than 350 ppm—a value we have already exceeded. Today's CO_2 level of over 400 ppm should be a cause of alarm for governments and indeed us all; even though we are not currently seeing a 'pre-advanced civilisation' climate situation.

It is too late to cap atmospheric CO_2 at 350 ppm and it is very clear now that we should have pulled out all the stops in the early 1990's to cap it at that level. Today, it will be a much better outcome if we can cap CO_2 at 420 ppm rather than let it run out to 450 ppm.

---

[25] 400 ppm - 300 ppm/300 ppm * 100 = approx. 33%

Not only but Also

The average global temperature has, so far, increased by just over 1½°Fahrenheit during the industrial era, with most of the increase occurring towards the end of the 20th and start of the 21st century—this increase is between 0.8°C and 0.9°C. I contend that it is quite possible that the planet has not yet warmed to a level that 400 ppm CO_2 might indicate. Although CO_2 has increased by 43%[26], the mid-point temperature rise (of the above range) is only 0.85°C. This represents, in absolute terms, a rise of just over 0.3% (Kelvin scale of temperature[27]).

This is the uncertain bit; there is plenty of room for future tipping points. This is not to suggest that humanity is looking at an increase in global temperature of 43%. The increased infra-red radiation, emitted from a hotter Earth, would stabilise the temperature long before a +43% temperature level is reached. However, no one can give an incontrovertible guarantee that even the much publicised increase in average global temperature of 2°C will be safe.

From where we are in 2015, this is very supportive of a position that there is no room for complacency in respect to setting emission reduction time-lines. If humanity is indeed being given a lifeline of the temperature rise lagging behind increases in CO_2 emissions, it will be thanks to the size of the Earth. With a system as large as our planet, with a mass of 5.98×10^{24} kilogram, there is bound to be a lag between cause and effect.

Hysteresis is a term that describes the time-lag between the occurrence of an event and the response of a controlled, or uncontrolled, system. In a planetary system, the time-lag would be significant. It is possible that the reason we are not

---

[26] CO_2 is measured automatically in absolute terms as the horizontal axis is positioned at zero.

[27] A one degree graduation on the absolute temperature scale, Kelvin, is equal to a one degree graduation on the Celsius temperature scale – it is the zero setting that differs. Zero Kelvin equals minus 273.6 degrees Celsius.

beginning to experience more extreme environmental effects right now, is the extreme speed at which atmospheric CO_2 is rising and the response of the natural world—planet Earth— cannot match that speed. But the danger is that the planet will catch up with our current and future changes to the chemistry of the atmosphere. The cost to our ordered society will be large.

Mankind is not playing for pennies with the Earth's climate.

The relative long-term stability of the highs and lows of CO_2 and temperature in the climate cycle, over several hundreds of thousands of years, has clearly been disrupted by our reliance on fossil fuels to power the industrial age. There is no excuse for continuing this practice now we are cognisant of the risks we are taking. As previously mentioned we now have the highest atmospheric CO_2 for at least 800,000 years, with some estimates of the highest level for two, three and even five million years.

Where are we supposed to gain comfort from these statistics?
Everyone I know prefers the climate we enjoy now.

Yes, 0.012% is Significant

As discussed earlier, global warming sceptics are fond of making mileage out of the numerically small increase in atmospheric CO_2 over the Industrial Era. They always compare the increase in the single gas (CO_2) to the total of all of the atmospheric gases. Today atmospheric CO_2 is about 400 ppm (412 ppm: May 2018) compared to pre-industrial CO_2 of about 280 ppm. This increase in atmospheric CO_2 (120 ppm) represents an increase of about 43%, is dismissed as inconsequential by sceptics as it represents only a 0.012% change in the total atmosphere over the Industrial Era.

It is flawed logic to compare the increase in CO_2 to the total atmosphere (100%). Consequently, focussing on the apparent minuteness of this figure is meaningless because the total pre-industrial CO_2 level was only 0.028% of the dry atmosphere in the first place.

Other than the major atmospheric gases (nitrogen, oxygen and argon), all of the remaining atmospheric gases add up to only 0.0422%[28] or 422 parts per million. In other words, or more correctly in fractional notation, *less than* **one half** *of one tenth of one percent* of the Earth's atmosphere contains all of the greenhouse gases, including carbon dioxide, methane, nitrous oxide, the chlorofluorocarbons and hydro-fluorocarbons and every other atmospheric gas on the periodic table.

The really scary thing is that in 1765 the figure was closer to **one quarter** *of one tenth of one percent!*

The 43% increase in atmospheric CO_2 is therefore a massive change to the chemistry of the atmosphere because CO_2 is the principal greenhouse gas, making up nearly 99.5% of the total atmospheric greenhouse gases. So, to reinforce the above, if carbon dioxide is included with the previously noted nitrogen, oxygen and argon, the remaining 0.0022%[29] of the atmosphere contains all the remaining greenhouse gases[30] and every other atmospheric gas on the periodic table.

So even though CO_2 is a tiny part of total atmospheric gases, it has been just right to maintain the world average temperature at that 'sweet spot' of about 14°C, which has enabled human civilization to thrive and prosper.

Switching now to water vapour, those denying human influence on global warming often assert that because water vapour is both a dominant greenhouse gas[31] and a natural phenomenon, then the 43% increase in CO_2 over the

---

[28] Calculated as 100% - 99.9578% (nitrogen, oxygen and argon) = 0.0422%.

[29] Calculated as 100% - 99.9578% (nitrogen, oxygen and argon) – 0.0400% (CO_2) = 0.0022%.

[30] These include methane, nitrous oxide, chlorofluorocarbons and hydro-fluorocarbons.

[31] Water vapour varies between a trace and 4% by volume, depending on both air temperature and location on the planet.

Industrial Era is of no consequence in comparison. Nothing could be further from the truth; a 43% change in any parameter is always significant.

Water vapour does have a significant greenhouse effect and together with the other greenhouse gases, is instrumental in keeping the planet at a comfortable average temperature of 14°C. But, it is not directly responsible for increasing the temperature of the planet; the phenomenon known as global warming.

Actions by humans have very little direct influence on planetary water vapour levels, but there is a danger that a significant rise in global temperature, because of emissions from industrial and agricultural activity, may enhance the ability of the atmosphere to hold water vapour. In turn, this may lead to a future increase in the greenhouse effect[32] and is just one of the potential global warming feedback mechanisms that should concern us all.

Natural versus Enhanced

The key issue in any debate on the relative influence of various greenhouse gases should not be about the gases that have been naturally occurring for hundreds of millennia. It is not these that are upsetting the delicate balance. The factor upsetting the vitally important balance in the atmosphere is global industrialisation and the attendant extra greenhouse gases that are giving rise to what is called the *'enhanced greenhouse effect'*.

There is an important but fine distinction between the natural greenhouse effect, which warms the planet and is essential to keep every one of us alive and the enhanced greenhouse effect that is upsetting the delicate balance in the atmosphere and causing global warming.

---

[32] Consequences of altering the water vapour carrying capacity of the atmosphere may trigger unknown and unfavourable tipping points.

The 'natural greenhouse effect' is sufficient
—we do not need to add to it.

A pharmaceutical analogy is useful. With the correct dosage of a drug or medicine, great therapeutic outcomes are achieved, but with excessive doses, the results can be very damaging or even fatal. It is not too different with greenhouse gases in the atmosphere. Our continuing involvement in altering such a system, before fully understanding all the intricate interactions, is naive to say the least. So far, the biosphere has absorbed the 43% increase in atmospheric CO_2 since the beginning of the Industrial Era, without significantly affecting human society, but with continuing high emission levels this cannot be assured for all time.

The ability of the atmosphere and oceans, to continually absorb greenhouse gases without causing excessive atmospheric heating and further acidification of the oceans is not limitless. Much of the natural absorption capacity is taken up by naturally occurring emissions such as decaying organic material, forest fires, volcanoes and methane from wild animals. These have been part of the natural cycle and in balance, for much longer than the period of modern human development.

Unfortunately, there is still work to do in getting the message of the danger associated with global warming across to the population at large and the people that govern us. There is little evidence of action to mitigate the increase in atmospheric greenhouse gases on a scale commensurate with the problem.

It should be clear to us all, even to the most trenchant climate change denialist[33], that there is an abundance of highly rigorous scientific evidence supporting the reality that the planet is warming and that this warming is a consequence of our industrial, agricultural and land use practices. Reports that

---

[33] Various terms to describe people on both sides of the 'global warming' debate are discussed in Chapter 4.

comprehensively document the Earth's warming trend are available from hundreds of well-credentialed scientific research organisations, Universities and the United Nations.

A vast pool of research items is readily available and this information is at the fingertips of everyone with access to the Internet[34].

Volcanoes

Volcanoes are worthy of special mention in the context of global warming. The denial side of the debate pounce on these spectacular eruptions suggesting that, for example, there is no point in downsizing the automobile fleet when a volcano can spew out more CO_2 in an hour than an automobile does in decades. This is deliberately misleading.

There are about 50 volcanic eruptions each year and the CO_2 from these volcanoes is included in the Earth's natural emission total. Even the most spectacular eruptions *are included in the natural emission total.* The 2010 eruption of Eyjafjallajökull in Iceland, which caused such disruption to European air travel, is part of the natural emission total, as is the 2011 eruption of Puyehue-Cordón Caulle in Chile, which sent ash plumes eastwards grounding aircraft in parts of the Southern Hemisphere.

Of course, there are potential events in nature that would be significant enough to derail our best environmental efforts. A giant asteroid smashing into the Earth is one and a 'super volcano', such as the one that erupted in Yellowstone National Park nearly 640,000 years ago, is another. However, mankind has not previously let the chance of a random catastrophic event paralyse society into inaction and should not do so now.

---

[34] The Internet is available in most Public Libraries, if no other connection is accessible.

Global warming is different as it is not a random event. It is a man-made potential disaster, only needing sufficient time to fully develop. This is the ultimate test of our determination, science and engineering. Action to develop solutions should commence without further delay.

It's Time to Act

Achieving a reduction in greenhouse gases on a scale necessary to stabilise the atmosphere, will require *everyone* to be working for the common good. This especially includes people in positions of power and influence, particularly those in the political and media arena.

Anyone with an ability to solve very complex problems will need to be recruited to this cause and this will certainly include the best thinkers in the behavioural and physical sciences and all branches of engineering. To successfully manage Planet Earth into the future, scientists and engineers should become among the most highly sought after of all the professions.

And that Time is Now

It is possible that a remedy to the atmospheric situation could be relatively straightforward if there were one hundred, fifty or even thirty years left to do the work. Scientific and engineering research could be allowed to follow a more natural course of events. More urgency is now required and we continue with the status quo at our peril.

The Earth's atmosphere is altering and the timeframes to critical tipping points are unknown. The time left from now to some future critical overbalance is rapidly diminishing. Effective changes to manufacturing practices and also changes in how we use our natural resources, minerals, energy/fuel, forestry and the atmosphere are currently minimal. Although our society appears to be slowly working to some kind of timeframe, there is no specific target on methodology, or timing.

The key challenges are time and change and the two are strongly linked. This book explores both topics and identifies solutions that must be linked to timely change. Time and change, where the time available is indeterminate and the change component may become more than we can manage, within any timeframe, if we are too tardy with the starting date.

An excellent example of anticipating and managing a potentially catastrophic event, with an uncertain timeframe, is the expenditure of many hundreds of millions of dollars to seismically retrofit the Golden Gate Bridge to withstand a magnitude 8.3 earthquake. Of course, it is to be hoped that an earthquake of such magnitude never occurs in the Bay Area of San Francisco, but it is better to be prepared and never be tested, than found to be wanting in a moment of crisis. The bridge retrofit is made even more difficult because not only could a seismic event happen at any time, but also the structural integrity of the bridge must be maintained during the entire retrofit process, enabling traffic flow to continue uninterrupted.

As with this example, which all would consider prudent, the timing of any potentially catastrophic climate event cannot be accurately predicted. We should all adopt a 'Golden Gate Bridge' approach to the changes that our industrial and agricultural practices are causing in the biosphere.

A difficulty for science may be that it is expected to make definitive and detailed analyses at all times. There is foolishness in expecting that anyone, regardless of impeccable academic credentials, could be totally precise in matters involving the biosphere of a 5.98×10^{24} kg planet. We should not and cannot, use this as an excuse for inaction.

Science has done its initial work and we are all well-informed of the dangers we face. Scientists cannot be expected to predict every aspect of what happens to the climate now, nor in one, two, or forty years hence. The weather bureau often does not predict even today's weather with 100% accuracy—but we still pay attention to weather reports!

So while not all parts of every ice sheet or snowfield will melt to the same extent, or necessarily to a greater extent each summer, action is still needed—despite denialists exploiting this natural variance with great zeal.

The 'canary in the mine' is ever-present and visible in the melting of polar ice, ice sheets, glaciers and the retreat of snowlines. Physical phenomena do not lie.

3. Complexity

Other than the expressed view of the most vociferous climate change denialist, most people across a broad spectrum of the community accept the overwhelming scientific data, satellite images, photographic records and their own observations and experiences, as evidence that the climate of the Earth is indeed changing. Currently the main directly measurable effects are a rise in average global temperature of nearly 0.9°C (just over 1½°F) and an increase in the frequency of 'extreme' weather events. The topic of extreme weather is further discussed in Chapter 5: The Weather.

Global warming and climate change are the terms most frequently used to describe the atmospheric phenomena increasingly presenting challenges to human society. While, both terms adequately describe the *currently observed consequences* of higher levels of atmospheric CO_2, neither term is adequate to describe the *likely future consequences* of the phenomenon. The terminologies global warming and climate change serve to deflect attention away from the underlying cause of the changes we are witnessing—namely CO_2. This focus on symptom, rather than cause, means that neither term is helpful in galvanising the world community into significantly reducing atmospheric CO_2.

Emission and Temperature Targets

From my reading, listening and watching, the overwhelming bulk of evidence indicates that the underlying cause of global warming is excessive emission of CO_2 into the atmosphere. Despite this, the notion that the variable to be controlled is temperature, rather than the absolute level of carbon dioxide (CO_2) in the atmosphere is reinforced at multiple conferences and press releases.

As previously mentioned there are widely accepted/agreed international commitments to limit the average global surface temperature rise to 2°C. However, there are no internationally agreed *legally binding* targets that will limit the amount of CO_2 emitted into the atmosphere. To achieve an actual limit on atmospheric CO_2 the world will need to limit the amount of carbon dioxide, or more precisely the amount of carbon[35] in tonnes, that can be emitted to the atmosphere.

It must still be determined if 450 ppm and 2°C are indeed safe limits or whether it should be closer to 350/400 ppm and 1°C to 1.5°C. The determination of accurate safe limits is urgent. The task of obtaining legally binding emission levels, fair to all countries, remains substantially within the purview of future international meetings and conferences. This is a heavy responsibility for delegates. Agreement on these matters will not be easy.

With the current temperature rise (above the pre-industrial level) approaching 0.9°C and a CO_2 concentration of above 400 ppm the lack of urgent political and economic action to reduce the rise in Earth's greenhouse gases is remarkable. It is hard to imagine any priority more important than the survival of our civilisation.

Flawed Logic

To illustrate the flawed logic of hoping to achieve a target temperature (the effect), without protocols for controlling atmospheric CO_2 (the cause), we can study the methodology of the simple x–y graph. As previously mentioned in this text, strict and legally binding protocols for controlling the increase of atmospheric CO_2 have been missing from all major climate change conferences.

Applying the terminology of a traditional x–y graph, the 'independent variable' is graphed along the 'x' axis and the 'dependant variable' on the 'y' axis.

---

[35] 1 tonne of carbon (C) equals 3.67 tonnes of carbon dioxide (CO_2).

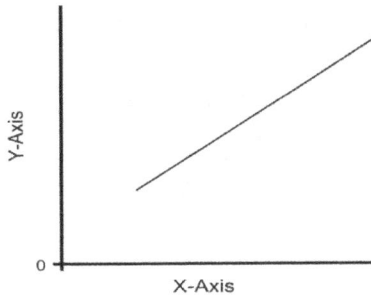

In the case of CO_2 emissions, the emissions—the independent variable—would be plotted on the x axis and the dependant variable, the global temperature, plotted on the y axis. The controlling factor is always the independent variable.

The following graph illustrates the fundamental flaw in attempting to arrive at a value for a dependant variable without specifying the independent variable. This is a graph of average monthly rainfall to the month in which it occurred.

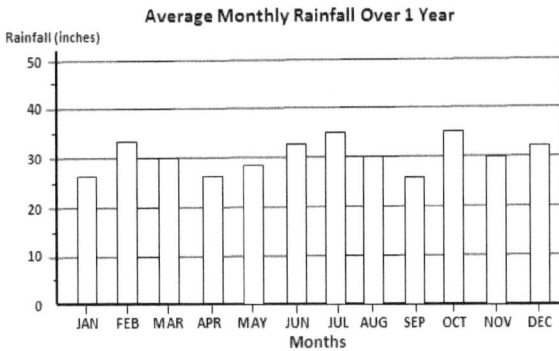

In this example, the independent variable is the month (x—horizontal axis) and the dependant variable is the average rainfall (y—vertical axis). While it is clearly possible to determine the average rainfall for a particular month, it is obviously impossible to determine the exact month in which the average rainfall was, for example, 30 inches, because there were three different months where the average rainfall was 30 inches. The question 'in which month was the rainfall 30 inches?' cannot produce a single definitive answer.

In the global warming context, it is not logical to specify a target for the dependant variable (temperature), because the variable to be controlled is the independent variable, namely CO_2.

Setting a target for reducing global temperature rise, without controlling CO_2 emissions, is akin to setting a target for personal weight loss without controlling either exercise levels or kilojoule intake!

Aiming to limit the increase in average global temperature can only be an aspirational target, at best. Future average world temperature will depend on controlling the overall concentration of man-made greenhouse gases in the atmosphere, with CO_2 being the main individual gas.

It is crucial therefore, to set a binding target for the maximum level of CO_2 in the atmosphere. Agreement on this is essential to guarantee that our way of life continues uninterrupted, as was nearly achieved in 1992 with the Kyoto Protocol. The aims of this protocol should have been strengthened and not marginalised, as has subsequently occurred. When binding targets for CO_2 emissions are agreed, an enforceable price for using the atmosphere as a CO_2 dump should apply to all nations.

This will be the start of the long battle to protect our habitat.

One Trillion Tonnes

Up to this point, we have talked about parts per million carbon dioxide (CO_2), or parts per million carbon dioxide equivalent (CO_2e), but not carbon (C) the element. Parts per million (ppm) of *both CO2 and CO2e refer to the measured atmospheric outcomes,* which are the *result of natural and anthropogenic emissions of greenhouse gases.*

To place limits on anthropogenic emissions, the measure used is not parts per million (ppm) of CO_2 in the atmosphere, but the *actual mass of carbon, or carbon dioxide,* in tonnes that is *emitted into the atmosphere.* Sometimes the measured quantity is specified in terms of tonnes of carbon[36] and other times in tonnes of carbon dioxide. Both conventions—tonnes of carbon (C) and tonnes of carbon dioxide (CO_2)—will occur in the media depending on the originator of the data.

It has been frequently reported that to keep global warming below 2°C, all-time carbon emissions, cumulative from the start of the industrial era to forever, must be kept below about 1,000 billion tonnes[37] (3,670 billion tonnes CO_2). It must be remembered that a 2°C temperature rise is not a universally agreed target and there are commentators who point to 1°C to 1.5°C as a maximum, reasonably safe, temperature rise.

To achieve the lower target of 1°C, all-time carbon emissions should probably not have exceeded 500 billion tonnes. As the total emissions of carbon (C) into the atmosphere are already approaching 600 billion tonnes and CO_2 is at 400 ppm, the limits of 500 billion tonnes of carbon and 350 ppm atmospheric CO_2 have already been passed.

As a world community, we must now do everything possible to keep total carbon emitted into the atmosphere well under 1,000 billion tonnes. Unquestionably, 750 billion tonnes would be a preferable target, a target that may give us a chance of limiting the temperature rise to 1.5°C.

---

[36] To convert a quantity of carbon (C) to carbon dioxide (CO_2), multiply by a factor of 3.67.

[37] 1000 billion tonnes equals 1 trillion tonnes.

Where are we now?

These days we switch on the radio or television and often hear that the concentration (ppm) of atmospheric CO_2 has risen yet again. Being aware that part of the emissions are still being absorbed by the oceans, it is clear that the annual tonnage of carbon/carbon dioxide (C/CO_2) emitted to the atmosphere is increasing rather than diminishing.

If total global emissions continue to increase at the rate they are, the 'trillion-tonnes' target could be passed by 2050. However, if 750 billion tonnes of carbon is indeed the safer limit, this could possibly be exceeded around 2030.

During the next two decades there is scope for the above projections to vary in either direction depending on our collective action. Of course, we all hope for that variation to be in a direction that benefits society. If our current model of 'business as usual' is continued as the prevailing economic paradigm, it is more likely than not, that any variance will be relatively minor in comparison to the problem of global warming.

Where are we going?

The complexity of climate change is heightened by several factors. It is of concern that nearly 600 billion tonnes, of a possible maximum of one trillion tonnes of carbon, is already in the atmosphere.

Further, the demand for electricity, which is currently very carbon intensive, is expected to at least double by 2050. Adding to this situation, the need for fuels/energy[38] for uses, other than the energy required to generate electricity, will also increase. In addition, the total global energy demand at any date in the future will depend on the Earth's population at the time and the standard of living of that population. Given this, a doubling in demand for both electricity and other

---

[38] Usually measured in millions of tonnes of oil equivalent (Mtoe).

fuel/energy types (from 2015–2050) may be considered a very conservative figure.

We do know, however, that even today's energy requirement, for all purposes, is enormous. We do know that Earth is a massive planet with a very large population and an insatiable appetite for energy. Our challenge is to meet the demand.

A long-term solution to the world's energy demand will require more than a continuation of the current 'business as usual' economic model. Chapter 12: Science and Engineering – the Future, expands on the notion of extreme science, as well as other future technologies.

As land transport is transitioned to electric and fuel cell vehicles and the world moves to solar-thermal energy sources for electricity generation, the reliance on fossil fuels will be reduced. This should lead to a corresponding reduction in the total global energy required, in percentage terms, to cater for the increased demand for electricity. Depending on the generation method, electricity is usually environmentally superior and is also a far more efficient energy source.

Estimated increases in electricity demand alone, does not tell the full story. The percentage of power generated from nuclear, solar and other renewable sources, in addition to the contribution from energy efficiency measures will be pivotal.

At this time, any new methods of electricity generation and enhanced energy efficiency measures are largely undefined. Currently, less than 20% of energy/fuel requirements are being met by non-fossil fuel sources of energy. These sources include nuclear, hydro, wind, solar, tidal and bio-fuels.

Bio-fuels are a special case. They are often considered neutral for the environment on the basis that the CO_2 emitted during the combustion phase is re-absorbed during the growing part of the cycle. However, it is not entirely a 'zero sum game' because no thermodynamic process is 100% efficient.

Additionally, the competition between land usage for food production and land for algae farms/bio-fuel crops will almost certainly become problematic, with the world population expected to increase to over nine billion people by 2050. See Chapter 16: Sustainable Manufacture.

Also, in many parts of the world energy is still obtained directly from the burning of fossil fuels such as wood, charcoal and peat. This is part of the total global energy equation and could increase as the world population reaches the expected nine billion people.

Change

The climate is changing, but we are not. A new approach is needed to deal with the emerging crisis; this must include attitude, economic mindset and real action.

We can start by changing how we describe the problem. Rather than 'global warming' or 'climate change', a more apt term would be the *'emerging atmospheric crisis.'* We need change in our thinking and change at all levels of society. The change will be challenging and will need to be very carefully managed.

Starting right now, our society needs to do everything possible to halt and then reverse the rate of CO_2 emissions of anthropogenic origin, into the atmosphere. This is the essential first step in limiting all greenhouse gases. Changing the dominant discourse on CO_2 emissions is at the heart of this book; provision of a blueprint for setting a different course.

Changes in the chemical composition of the atmosphere must be cast as the villain. Advanced human civilization is part of the solution. While our part in these changes may have been unwitting in the past, this is no longer the case.
We now know the reality. Furthermore, it must always be remembered that global warming and the changing climate are the symptoms—not the cause.

Change must include the whole world community: government, business, industry and individuals. Everyone in society will need to contribute appropriately to the capital and ongoing expense of reducing CO_2 emissions. All of us will benefit from the ultimate gains.

Some works on global warming and climate change maintain that the technology required to fix the crisis already exists. I do not agree. To solve this crisis, the world needs substantial change to methods of energy generation and transmission, liquid fuel use and the manufacture and distribution of consumer goods.

Given the very real potential for disaster, a comprehensive range of proposals to reduce emissions should be fully evaluated. Additionally, the financial commitment to reduce atmospheric greenhouse gases must be massively increased.

Worldwide, government policies and priorities must increasingly focus on funding and facilitating innovative scientific and engineering research projects. In particular, projects that are specifically designed to accelerate the elimination of CO_2 emissions from the energy, fuel and manufacturing sectors of the economy should be fast tracked.

Planning and action ought to begin now on the best ways to achieve zero net emissions by 2030 from energy generation and transmission, transport[39] and from the manufacture and distribution of consumer goods.

Even though there is a need to do a great deal more, the importance of existing emission reduction measures and projects should not be undervalued. Also, any proposal for new emission trading systems, direct taxes on CO_2 emissions and other direct action emission reduction strategies must be seriously considered. Energy efficiency and emission reduction projects, already in existence, will all help to reduce

---

[39] The aviation sector is more fully discussed in Chapter 16: Sustainable Manufacture.

the rate of increase of atmospheric CO_2. Existing initiatives should continue while the new programs are brought on line.

The expansive and visionary projects required to extricate humanity from the circumstances into which it has accidentally stumbled, will require substantial funding—the big question is how much? The ongoing total of all government assistance to the financial sector, provided to tackle the financial crisis that started in 2008, would be a good start. This is discussed further in Chapter 12: Science and Engineering – The Future.

It is very important not to underestimate the seriousness of the situation that faced the world economy when global credit markets became dysfunctional. Decisive worldwide government action was urgent and appropriate to prevent collapse of the entire financial system. Equally, the emerging atmospheric crisis should not be underestimated. A full-on crisis in the atmosphere would be considerably worse than any financial crisis. Stabilising the chemical composition of the atmosphere must now be humanity's number one priority, just as the financial crisis was in 2008/09.

From Rhetoric to Reality

Everything involving the atmosphere or the politics of the atmosphere, is extraordinarily complex. Resolution of the political issues and finding solutions to the economic, scientific and engineering problems will be equally complex.

World leaders often make statements regarding global warming, climate change and other problems stemming from the elevated level of greenhouse gases in the atmosphere. Climate scientists are deeply concerned about this issue and some in the scientific community are making predictions that are ominous for the world community in the longer term, if action to stop the rise in atmospheric CO_2 and global warming is not taken very soon.

There are some in the scientific community that deny global warming and the likelihood that the world will face consequences from the burning of fossil fuels. However, these scientists are increasingly isolated from mainstream climate research and often have qualifications in disciplines unrelated to climate science.

Then, there is also the evidence of our own eyes, both from direct observation of local events and the photographic record of changes that are occurring around the world. Evidence of more geographically distant negative climate events is seen in the print and visual media almost weekly. As yet, we are not observing mitigation action on a scale commensurate with the vast amount of evidence that the planet is warming.

Political leaders have unequivocally outlined an emerging problem with greenhouse gases in numerous statements over many years. These include speeches dating back to the 1980's, when Britain's former Prime Minister, Baroness Margaret Thatcher, addressed the Royal Society[40] in 1988, the United Nations General Assembly in 1989 and the second World Climate Conference in 1990.

If any judgement on global warming were to be made based on these speeches, as well as statements from leaders of the G20 group of industrialised nations, the United Nations and influential leaders from the business world, it would be clear that we have a very serious situation, deteriorating at an escalating rate. These influential people and groups have access to significant and reliable scientific briefings and from what they are saying, it would appear they are being briefed on alarming scenarios for the world if the build-up of greenhouse gases in the atmosphere continues unchecked.

More than twenty five years after the first warnings, we now have even clearer indications of a developing dysfunction in the Earth's environment. We all know the environment is

---

[40] An extract from the 1988 address to the Royal Society is included in Chapter 6: Science and Politics - The Antithesis.

at risk. Everyone talks about it, both public figures and our own acquaintances. Yet, where is the action equal to the task of repairing the environment of a 5.98×10^{24} kilogram planet?

In fact, there is *an astonishing lack of action* on a scale sufficient to fix a planet-wide problem and to quickly move industry and commerce to a climate friendly—zero net anthropogenic CO_2 emission—economic model.

There is the observable division between the scientific community and society at large. Few topics polarise public opinion as dramatically as global warming and climate change and this separation is dangerous in that it stands in the way of achieving any consensus to action. Much of the public discourse is still focussed on whether climate change is actually happening. If the world's scientists are worried by the increase of greenhouse gases in the atmosphere, why is the population at large not similarly concerned?

The disparity may be explained by the fact that there are powerful elements within society which stand to benefit from a continuation of status quo economics and are able to disproportionately influence the media. The immediate interests of this group are best served by propagating the view that there is no connection between greenhouse gas levels in the atmosphere and CO_2 emission intensive industrialisation.

There is a seductive appeal to a message which is essentially, 'don't worry, nothing bad is going to happen and if something bad does happen, there's nothing we can do about it anyway; it's all part of a natural cycle.' This may be an appealing message, but it is not based on present climate science and is neither helpful nor constructive. With Western societies mollified by such misdirection[41] they are not demanding the necessary changes from their politicians.

---

[41] A favourite technique of the magician.

As noted in Chapter 2:

'With both consumers and business comfortable with the status quo, there is little incentive for global action on reducing emissions commensurate with the scale of the problem.'

Bring on the Future

We have collectively chosen and are continuing to choose a path towards full industrialisation of society based on the use of fossil fuels. There is nothing essentially wrong with a fully industrialised society; it can spread great benefit to even more of Earth's people; it is just the way we pursue it that needs to change.

The energy and manufacturing systems needed to fix the emerging atmospheric crisis have not yet been invented. At best, they may be in the early stages of theoretical research. Wide-ranging research and development is urgently required and must be adequately funded by both the government and private sectors. Definitive and practical engineering solutions[42] will follow the research and the research will follow the funding. Adequate funding will only be prioritised if enough people demand a change in public policy on the use of fossil fuels. Revised economic models will emerge from a political determination to do something significant about global warming.

It is impossible to know which of any new technology will form the backbone of a future zero net emission world economy. Attempting to predict which of any new discovery will deliver the required results challenges the whole premise on which original research is based.

---

[42] To name just a few: this includes branches of engineering such as civil, mechanical, manufacturing, electrical, power electrical, electronics, chemical and nanotechnology.

4. Cutting through Confusion

With complexity comes confusion. Uncertainty and confusion are understandable responses to any discussion about the Earth's climate. The point is frequently made that the average person cannot possibly understand every consequence of global warming and climate change. Those in the community who argue against taking action to reduce CO_2 emissions, often use the lack of understanding in the wider society as a lever for their argument.

For example, emission trading schemes and direct taxes on CO_2 emissions are complex concepts and it is difficult for the layperson to fully understand all aspects of the economic argument. This has been used with great effect to confuse the climate debate. The resulting uncertainty makes it much easier to argue a case for doing nothing.

The wider community finds it difficult to decide if an individual, a country, or indeed the world is doing too much, too little, or taking the correct measures to mitigate atmospheric CO_2 emissions. The lack of an effective agreement to limit CO_2 emissions into the atmosphere, at any of the recent conferences on climate change, is testimony to the power of the 'confuse and conquer' strategy.

There are unimaginable numbers of interconnected interactions continually occurring within the Earth's biosphere, of which the atmosphere is but one component. In complexity, the biosphere is the ecosystem equivalent of the 'many-body problem' as defined in quantum mechanics[43]. It is probably beyond our capability to accurately predict the effect of the chemical changes that are currently occurring in the composition of the atmosphere.

---

[43] The quantum mechanics term 'many-body problem' refers to microscopic particle systems containing very large, or infinite, number of particles that interact within the system.

We cannot use this complexity as a reason or excuse for delaying action towards preserving a state of equilibrium in the atmosphere. In fact, this complexity adds to the need for substantial and fundamental changes to our current economic and consumption paradigms.

Information Overload

One reason for confusion is that we are overwhelmed by information on matters of climate. Some of this information is conflicting, if not with other information within the climate debate, then with our own value systems.

A feature of the information age is that after burying us in information, it then provides the tools to selectively dig ourselves out—such as search engines, databases and online study materials. This is particularly helpful in matters of climate as any work attempting to incorporate all information pertinent to specifying environmental objectives out to 2030, would be unwieldy and of little use.

This is the reasoning as to why details of well-known concepts that are readily available in the public domain are not revisited in this book. Most readers will either possess sufficient knowledge, or can readily explore any concepts discussed should they wish.

Pseudo-religion

A tactic of those who deny the reality of global warming is to suggest that people calling for more action on climate change are part of a new pseudo-religion of climate. If the words of people opposing action on climate change are carefully considered, it is easier to conclude that their side of the discussion better fits the description of a pseudo-religion.

A central tenet of most religions is usually a faith in a Deity that does not have a physical presence in this world. The description of pseudo-religion elegantly fits a climate change denialist viewpoint, who by sheer faith, stay with a world-view that is also without evidence in the physical world.

False Comfort

A situation is developing where some people are, perhaps unintentionally, underestimating the gravity of climatic changes. This is exacerbated by the choice of language used in public debate. Frequently, euphemistic words, soothing phrases and simplistic examples are used in connection with climate issues. This is leading to a false sense of security in the community. The following examples are not definitive but demonstrate a trend in the language used in public debate.

A typical example may be a television news program or documentary. The program features a young person engaged in a local environmental project sponsored through a school, or similar organisation.

During the segment wrap-up, the program host may make a comment such as 'with such first-class young people concerned about the environment, the future of the planet is in good hands.' While not seeking to diminish the efforts of said young person, one must question this summary. The power to significantly alter the economics of energy, transport and agriculture, does not rest with twelve to sixteen year olds and by the time it does, it will be too late.

'Save the planet' is another favoured euphemism. Most of us know that 'the planet' will not mourn the absence of advanced human civilization. In truth, 'the planet' will start a healing process much sooner, in the absence of humans, than it will under our current stewardship. Saving civilization, otherwise known as 'our way of life', rather than 'saving the planet' is the real challenge facing every person on Earth today.

These examples are not important in their own right, except to reinforce the emergence of a trend that trivialises a very dangerous situation. Whether by intent or by accident the widespread use of such expressions and clichés serve no useful purpose.

It is important the language used in connection with the atmospheric crisis does not convey a false sense of security. The reality is confronting, but until the world community acknowledges the real issues that must be addressed, we cannot start to fully understand the challenges faced by us all.

Political leaders often suggest that we may be failing 'future generations' by not acting decisively to reduce CO_2 emissions. This also 'sugar coats' the problem. It is a gross misunderstanding of atmospheric science to believe that we can avoid significant and negative environmental consequences manifested well before the 2030's.

We shall also be failing current generations. Starting with the 'silent generation[44]', followed by 'baby boomers', generations 'X', 'Y', 'Z' and babies now being born[45], we already have six generations that will be affected by changes to the Earth's atmosphere; without even thinking about 'future generations'. This has relevance for everyone on planet Earth right now.

The notion of denying or deferring the impact of adverse outcomes from continued atmospheric change is not helpful. Humanity has no choice but to develop a realistic position about the changes we are causing to the Earth's atmosphere. If we find the political determination to take appropriate action, the Earth will continue to be our environmentally friendly home for a very long time.

Global people-power will be crucial in achieving this change to the political agenda.

---

[44] The 'silent generation': people born between the middle of the 1920s and 1946; the youngest of these folk will be 85 years old in 2030.

[45] Babies born from 2010 onwards are generally known as generation alpha (Gen A). Also, some demographers use 'The Millennials' for those born between the early 1980s and the early 2000s.

The role of people, whose only part in the political process is their vote, will be crucial in the battle to preserve the current environment.

What's in a Word?

There are many 'words' in the climate debate that add to the confusion. Words such as 'denialist', 'denier', 'sceptic' or 'skeptic,'[46] 'warmer' and 'warmist'[47], are used without qualification. The focus of any campaign to eliminate greenhouse gas emissions into the atmosphere is thus blunted. The focus should be on the fact that the level of atmospheric CO_2 grew by nearly 15 ppm between 2008 and 2015; that CO_2 in the atmosphere is now the highest it has been for hundreds if not thousands of millennia and that polar ice is melting at levels unprecedented in modern history. There is no confusion in these numbers and observations.

As previously discussed, depending on the source, there is close to a 2°C average global temperature rise which has already occurred, due to past and current emissions. It is widely assumed, but not universally agreed, that this temperature rise of 2°C and/or a CO_2 concentration of 450 ppm in the atmosphere are maximum safe levels. As discussed in Chapters 2 and 3, how anyone could know that these numbers are safe is difficult to comprehend. The 2°C and 450 ppm cannot be scientifically verified as safe for our complex civilization, because the world has simply never before been in such a situation.

---

[46] Denialist, denier, sceptic, skeptic are names given to those who believe the Earth's atmosphere and oceans are either not warming, or any warming is not caused by humans and as a consequence do not see a need for human intervention or action.

[47] Warmer and warmist are names given to those who believe the Earth's atmosphere and oceans are warming, that this is a result of human industrial CO_2 emissions and action is needed to limit emissions in order to halt the warming trend.

However, if it is accepted that 2° Celsius (2°C) is really the safe limit and if it is also accepted that this limit actually does equate to 450 ppm CO_2 in the atmosphere, then we shall also be accepting a CO_2 level that is 60% above the pre-industrial level of 280 ppm. As the current level is already 43% above this pre-industrial level, we should all be aware that the factor of safety is diminishing rapidly.

It is time to focus on the real issue and that is whether or not humans are in agreement with increasing atmospheric carbon dioxide by 50%, 60% or even further above the pre-industrial level of 280 ppm. These elevated levels will certainly be much higher than anything that has prevailed on Earth, for a very long time! My vote is for us not to challenge this particular world record.

So, for the time being, if 'denialist', 'denier', 'sceptic', 'skeptic', 'warmer' and 'warmist' are defining the global warming debate, so be it! But let there be no confusion about the real issue facing us all.

5. The Weather

The weather is capricious by nature and a distraction from the real issue. Walking the High Street on a cold and wet day, feeling very much at one with the stereotypical frog in the icy pool, we may be greeted by someone saying 'so much for global warming!' The crisis in the atmosphere is frequently distilled down to local temperature, rainfall or put simply— the weather. High Street on a cold, wet day is not the place to engage in a discussion on the weather being a sub-set of the climate overall.

This spawns difficult questions such as 'why, in the winter of 2010 and 2011, did Britain record the coldest spells since 1981 and depending on the set of figures referenced, prior to that, since records began in 1910?' Also, 'why did 2009 see Perth, the State capital of Western Australia, record its coldest November since 1971?' 'Why did south-eastern Australia experience record-breaking low rainfalls in the years leading up to 2010 and then extreme wet conditions in the winter and spring of 2010 and 2011?'

At the same time as a small number of record-breaking 'cold and wet' events are occurring, many more parts of the world, including my home country of Australia, have experienced record higher temperatures. All around the world, the climate is changing and getting hotter.

Since the middle of the 20th century, Australian temperatures have, on average, risen by about 1°C with an increase in the frequency of heatwaves and a decrease in the numbers of frosts and cold days. Five years ago, winter temperatures in Canada were up by nearly 1½°C and in Greenland, by nearly 2½°C. The unseasonably warm and wet weather in the Vancouver area (Cypress Mountain) was a major cause of concern for the organisers of the 2010 Vancouver Winter Olympics in the weeks leading up to the games. Only a late 'cold snap' and snowmaking machines saved the day.

It is said that nine of the ten warmest years on record have occurred in this century and the other one only just missed out (1998), because it was at the very end of the 20th century. By all accounts, 2014 has the prize as the hottest year on record so far. The common link between the above examples and other anomalous weather events are changes to the chemistry and temperature of the atmosphere and oceans. In the case of the world's oceans, in addition to temperature and chemical changes, there is also thermal expansion and run-off from melting land ice, both of which are causing a rise in sea-level.

In the arctic regions a further issue is a reduction in the volume/area of the ice cap. This leads to changes in the arctic air mass above the ocean; up to and including the stratosphere. As detailed later in this chapter, there is a link between record-breaking weather events and global warming.

Record-breaking hot days exceed record-breaking cold days by a factor of between ten and twenty. Further, a large number of record cold temperatures that were set early in the last century still stand, but most of the record-breaking high temperatures have been in the last thirty years.

The Arctic

Extreme weather is occurring more frequently around the world. In January 2014, the weakening westerly winds in the polar jet stream may have caused changes to the pattern of the polar vortex that brought extreme cold to the north-eastern and central regions of the United States. In October 2012, the extremely unusual track of Superstorm Sandy, which also affected the north-eastern United States, was attributed in part to a weakening of the polar jet stream and a northward 'bulge' of the polar jet stream in the North Atlantic. There is evidence to suggest that both of these events can be linked to global warming. Relative to the rest of the Northern Hemisphere, there is an accelerated warming of the stratosphere at high latitudes in the Arctic regions.

The weakening and destabilising of the high latitude, high altitude, high wind-speed, polar jet stream causes it to break out of its usual oval-shaped path and 'loop' down to much lower latitudes. This allows extremely cold polar air much further south than would otherwise be normal. This phenomenon was relatively rare in the past, but with the rapid warming of the Arctic, the polar jet stream is weakened more often and tends to 'meander' to the south more frequently.

Global warming is more pronounced in the northern Polar Region

As parts of the polar jet stream loop south corresponding 'bulges' of warmer air move northwards in other regions. Simultaneously, with the extreme cold in the north-eastern parts of the United States, northwest Canada and Alaska experienced unseasonably warm conditions.

Not all aspects of Superstorm Sandy were attributable to global warming, but the world-wide rise in sea-level, extra energy and moisture from higher than normal ocean surface temperatures and a large wave in the polar jet stream all contributed to the extra-tropical cyclone (Sandy) being unusually destructive.

As with everything linked to a 5.98×10^{24} kilogram planet, the physics of the atmosphere and oceans is multifaceted and complex. Our understanding of these branches of science owes much to the work of Dr Jennifer Francis[48].

Superstorm Sandy is an example of a highly complex atmospheric and oceanic situation.
The simple, one-dimensional messages of those who deny anthropogenic climate change are disingenuous and dangerous.

---

[48] Dr Jennifer Francis: Research Professor, Department of Marine and Coastal Sciences, Rutgers University, New Jersey, United States of America.

The World-over

The weakening and increasingly unpredictable path of the polar jet stream is not confined to the United States. Europe and in particular, the United Kingdom, have recently experienced increasingly frequent bouts of extremely cold Arctic-like weather. This is likewise attributable to a weakening of the polar jet stream, causing it to depart from its normal path. The arctic cold is then experienced further south than would be expected.

As ocean temperatures increase, the warmer water has the potential to transfer more energy into tropical and extra-tropical storms. As the internal energy of a storm increases, the potential for it to become cyclonic is also increased and the likelihood of it becoming a category four or five storm is also increased.

Cyclones, hurricanes and typhoons are all the same weather phenomenon with different names depending on the geographic location. Over the past 40 years, the numbers of category four and five storms have increased, mirroring the increase in ocean temperature.

Should we go Category Six?

In November 2013, Typhoon Haiyan caused immense devastation in the Philippines. It has been called a 'Super' Typhoon because the highest storm category is currently five, but this storm may well have been a category six—if such a category existed. Depending on the storm classification system in use, a tropical cyclone/typhoon needs only to have a central pressure of less than (approximately) 930 hectopascals (hPa) to be rated as a category five.

If the scale were to be extrapolated to a hypothetical category six tropical cyclone, it is quite likely that the central pressure would be specified as needing to be less than 910 hPa. Tropical Typhoon Haiyan came in with a central pressure of under 900 hPa. Less than eighteen months later, tropical Cyclone Pam devastated Vanuatu with a central pressure reportedly little more than 890 hPa.

Maximum and sustained wind speeds are also important in cyclone classification; wind speed in cyclones is directly related to the measured minimum central pressure.

The critical point to be made is that before the consequences of the changing climate began to be manifested, cyclones with the intensity of Haiyan and Pam were extremely rare in that part of the world. Now, weather events such as Superstorm Sandy and cyclones Haiyan and Pam are becoming more frequent.

If the world community does not take sensible measures to correctly categorise the increasing number of severe weather events, we shall have no rigorous method of judging the worsening climate situation, because all 'super' typhoons and hurricanes will be recorded as category five.

Yet More

There is an ongoing drought in the USA State of California which is being reported as one of the worst on record and quite possibly, the worst drought in that State for 100 years.

In 2013, on the opposite side of the Pacific Ocean, Australia experienced the hottest year since records began in 1910. In October of the same year, the Blue Mountains region of the Australian State of New South Wales faced a very serious bush fire outbreak. The intensity of this fire and the extreme October temperatures were both unprecedented.

In Australia, October is two months before the start of the summer season in December. Fires of this nature, in this region, are more likely to occur at the height of the summer bush fire season, in late January/February.

The significant numbers of aberrant and extreme weather events occurring all around the world are not figments of the imagination that reside only in the minds of climate scientists. They are real; they are happening; they are serious. They demand our attention and engagement with what is becoming a very unsafe situation. Record-breaking heatwaves, droughts, bush and wild fires are all signs that the effects of a warming planet are becoming manifest in our world.

The fluidity of the situation means the task of cataloguing all events is never-ending.

With all these examples—each generating many questions and arguments—how is any sense to be made of seemingly conflicting warming and cooling events? If the world is warming, counter-intuitive events, although considerably fewer in number than 'hot' events, will naturally generate confusion.

It's too easy to say: 'It's the weather, capricious by nature and most likely, whatever happens has happened before.' Although there is no single definitive answer, extreme cold or wet-weather events, do not negate the longer-term trend of a warming planet.

In broad terms, climate can be thought of as weather averaged over a long period of time, with the weather being the prevailing temperature and precipitation conditions over a time-span of hours, days, but generally less than two weeks. In the context of global warming, it is the temperature that is of most interest.

Average temperatures over thirty year time spans are frequently used in analyses of climate variation, with some more recent climate studies involving ten year time spans. Both weather and climate vary according to the region of the Earth being considered and are inherently complex fields of study. Every aspect of the mechanism that caused those unusually severe winter conditions in the United States, Britain and Northern Europe in recent years may not be completely understood.

Although it seems counter-intuitive to suppose that unusual cold is linked to global warming, it is arguable that the weakening polar jet stream is causing the extreme cold snaps of 'arctic-like' weather in some regions of the northern hemisphere.

The severe winter conditions in Europe may also be partly a result of a reported weakening in the strength of the Gulf Stream. It is well known that historically, the warming effect of the Gulf Stream is the principal reason for the mild winter climate in Britain, compared with much colder places of similar latitude.

It has been suggested that the reduction in Gulf Stream strength may be due to an increase in fresh water flowing into the North Atlantic caused by melting of the Greenland ice sheet. As the returning Gulf Stream passes south of Greenland, this fresh water forces it to a greater depth and changes the balance that has existed for a very long time. Or, it could be due to a change in wind patterns relative to the Polar Regions. Regardless, one thing is certain:

Whilst we seek a definitive answer for each and every detail of the climate argument, the polar ice and the glaciers continue to melt.

Glaciers are melting, in the Himalayas and in the Alps. Even though the Himalayan glaciers may not completely disappear by any particular date, *they are still melting* and attempts by people to discredit the environmental movement over this issue will not change this fact.

Polar Region ice loss is characterised by both a reduction in the area covered by ice and *more importantly, by a widespread reduction in the thickness of the ice.* The combination of a reduction in area and thickness, leads to a significant overall reduction in the total volume of ice caps and ice sheets.

The reduction in the volume of ice in the Arctic, Antarctic and Greenland is of great concern. The result of the changes we are making to the chemical composition of the atmosphere is already evident. The widespread effect on the planet's permanent ice sheets is staggering. Permanent that is until the late 20th century.

We are even talking of the possibility of a 2° Celsius rise in average global surface temperature and an atmospheric CO_2 level of 450 ppm—as if it will be no more of an issue than increasing the size of a 'take-away' meal.

The ice melt from the Polar Regions has far-reaching consequences for climate systems and is the strongest indication that we are observing of an unstoppable move towards a warmer and increasingly unstable planet. This topic is further discussed in Chapter 8: Melting Ice.

As a people, we appear to have great difficulty in comprehending the perils associated with destabilising an immensely complex system such as the biosphere and its atmospheric subset.

Southern Hemisphere

In the southern hemisphere, an area familiar to me is the south-eastern corner of Australia. This geographic area experienced almost fourteen years of drought from 1995 to 2009. Droughts in eastern Australia are usually coupled with low or negative values of the Southern Oscillation Index, commonly known as El Niño events. However, during the above period, the low rainfall persisted even during positive values of the index, which are commonly known as La Niña events. Until this particularly long drought, La Niña events usually brought rainfall back to the long-term average.

However, rainfall patterns in southern Australia may also be affected by warming and cooling cycles in the Indian Ocean. Research at the University of New South Wales (UNSW) suggests that a phenomenon called the Indian Ocean

Dipole (IOD)[49] is a possible cause of an atypical Australian rainfall pattern 1995–2009. The IOD is characterised by cyclic warming and cooling of waters in opposing regions of the Indian Ocean.

The lack of moisture-bearing winds and rain to south-eastern parts of Australia during the 1995 to 2009 drought may have been caused, in part, by Indian Ocean current and temperature patterns, rather than solely by particular El Niño and La Niña temperature events in the Pacific Ocean. Of course, any deficit in rainfall would have been exacerbated by coinciding El Niño events.

The abnormally long drought in south-eastern Australia (1995–2009), persisted despite concurrent periods of La Niña in the Pacific Ocean. Simultaneously, there was also an unusually long duration of a particular IOD phase that was not conducive to rainfall in the region under discussion.

This long period of drought was then followed by unusually high rainfall which correlated with simultaneous rainfall inducing Indian Ocean Dipole and La Niña events.

Connected and Complicated

Returning to the Atlantic Ocean, any reduction in the strength of the Gulf Stream, flowing towards Europe, could possibly lead to other changes in Atlantic currents. Because ocean currents around the world are linked, it is possible that any change to warm current flows in the North Atlantic may also affect ocean currents in the South Atlantic and Indian Oceans.

Therefore, my observation is that it is possible that *the melting of the Greenland ice sheet* and the corresponding effect on

---

[49] Research conducted at the University of New South Wales (UNSW), Climate Change Research Centre, by Dr. Caroline Ummenhofer and Professor Matthew England, et al., February 2009. Findings detailed in a paper accepted for publication in the journal *Geophysical Review Letters*.

the Gulf Stream, if proven, could also be influencing ocean currents around the world, including the Indian Ocean.

As previously discussed, the Earth is massive—it has a mass of 5.98 x 10^{24} kilograms. The resulting physics of the atmosphere and the oceans is multifaceted and complex. This complexity is manifested in our weather all over the globe. Thus, rainfall patterns in south-east Australia may have much more complex origins than simply a La Niña or El Niño event.

Climate Science and the 'One in One-Hundred Years' Panacea

At the heart of this issue is that particular local weather events in Britain, Australia or any other part of the world, even if appearing to run counter to 'global warming', do not negate the overwhelming scientific evidence that shows a planet-wide trend towards a hotter future. Whether the weather is unseasonably dry, wet, cold or hot, the issue is that long-term seasonal weather trends and patterns are being disrupted.

In Australia, extreme weather is often explained in terms of one in ten, twenty, fifty, or one in 'one-hundred' year events. In recent times, the term 'a one in two-hundred year' weather event was even used! The idea of continually extending the time period, to maintain the appearance that these events are part of a natural cycle and thus avoid facing the unpleasant fact that we are changing the climate, is farcical. It grossly understates the underlying problem.

The underlying problem is that if weather events are described as the wettest, driest, hottest or coldest, for a certain number of years, they appear to be simple variations on weather themes that we have all seen before. There have always been floods, droughts, heatwaves and extreme cold. We all know and have experienced weather events that are often excused by statements such as 'but it's not really that much worse than occurred, for example, in 1974.'

An encouraging sign is that a few commentators are beginning to use the word 'unprecedented' in relation to some of the more extreme weather events. This is more realistic than describing extreme and aberrant weather as a 'one in five hundred years' event.

While *single weather events* may not be definitive proof of an atmosphere in crisis, more *frequently occurring random and aberrant weather events* are not as easily dismissed. It is the *aggregate of the separate events* that are currently occurring that is the clear departure from the long-term norm.

All of these individual weather events, taken together,
become the unprecedented event.

Although it is impossible to determine with certainty if unusual weather patterns in different parts of the world are linked to each other, let alone to specific atmospheric events, the likelihood of this being a coincidence ignores the glaring truth that a particular chemical constituent of the atmosphere, CO_2, has been increased by 43% compared to pre-industrial levels.

Around the world there are enough examples of changes to weather patterns, snow lines, glaciers, sea levels, droughts and other extreme weather events, which correlate in expected ways with science's predictions for a warming planet.

It is really only the exact mechanism of the linkage
that eludes our present science.

Therefore, the importance of establishing significant measures to substantially cut the amount of greenhouse gases that we release into the atmosphere cannot be overstated.

It is not necessary to seek a definitive answer for each and every detail of the effect on the atmosphere of increased greenhouse gas levels before taking such action. With each year that effective action on emissions is delayed, the inescapable fact is that even more polar and land-based ice is melting.

Ice melt is the unambiguous measure of a warming planet.

Interconnected

The difficulty in identifying exactly all of the complex interconnections between ocean currents, climate systems and anthropogenic CO_2 emissions has been demonstrated. To do so would require far more years of research than is available.

If the world is waiting for something catastrophic to happen that can be directly and unarguably linked to global warming, before taking action to preserve the biosphere, it will be too late. Preservation of our civilization demands that we take substantial action to stabilise the chemical composition of the atmosphere long before every last detail of such a complex system is known.

Making an analogy with human health, some of the illnesses that can end with a person's death often start with a relatively vague or minor symptom, ache or pain. Sometimes, these early indicators are ignored at our peril. So it is with the biosphere. The effects of global warming that are currently being experienced, are *being perceived as relatively benign* and as such, can be accommodated within our current economic paradigms. This will not always be the case.

Humanity *must focus on the 'big picture'* and we should not expect to reach significant 'big picture' conclusions from observations of short term, *'little picture' phenomena such as the 'weather'*. The more reliable indicator is the longer term climate phenomenon—remembering that the usual time span for quantifying climate trends is thirty years.

There is an important caveat to this. The 2015 calculation of climate will include temperature averages dating back to 1986. In a situation where the global temperature is rising, temperature averages from the early years of the thirty year cycle will tend to average down the influence of the later and hotter years, thus also averaging down the temperature component of the final climate calculation.

A Plethora of Indicators

Powerful and significant indicators of a warming planet are the number of times that record daily maximum temperatures have outnumbered record daily minimum temperatures over the past three decades. As stated earlier, record-breaking hot days are exceeding record-breaking cold days by a factor of between ten and twenty. A large number of record cold temperatures that were set early in the last century still stand, but most of the record-breaking high temperatures have been set at the end of the twentieth, or the start of the twenty-first century.

Stop press, July 2015: *Yet another exception that doesn't disprove the rule.* Much has recently been made of the southern Australian state of Victoria recording the coldest July for twenty years—but that is only back to 1995. As already discussed, there will be hot days and cold days, that is the nature of weather, but the overarching and inescapable reality is that the world's ice just keeps on melting.

It is not only in the temperate zones that heat records are being broken. In the Polar Regions, the signs are even more unmistakable.

A current and reliable indicator of the effect of higher levels of atmospheric CO_2 is the direct observation of the polar ice masses and what is happening to them.

The change in the Greenland ice sheet, Arctic sea ice and both sea ice and ice sheets from the Antarctic can be observed and measured, with considerable precision. There need be no confusion when it comes to matters of climate. The world will experience normal weather and extreme weather; it may get hotter, colder, wetter or drier in different parts at different times. Still, the planet's atmosphere is changing inexorably and nowhere is this more evident than in the Polar Regions.

Ultimately, success in the fight for civilization will be seen in the Polar Regions; the Arctic and the Antarctic, where mankind is scarcely represented. Perhaps this is another reason we are able to continue with *business as usual*, oblivious to the risks we are facing.

6. Science and Politics
— The Antithesis

Current global action to mitigate an emerging atmospheric crisis is inadequate. The message from scientists of many disciplines is unmistakable and yet it is substantially sidelined at a global political level. Individual politicians may make powerful statements that support the view that the science of global warming is indisputable and that climate change is real, but corresponding action is minimal.

Referring once again to the 1988 Royal Society speech by Britain's Prime Minister, Margaret Thatcher, in which she eloquently outlined the threat that global warming posed to the environment in general and the climatic system in particular, Prime Minister Thatcher's speech was insightful at the time and even more so, after the passage of twenty-seven years. The speech included the following[50]:

> *The danger of global warming is as yet unseen, but real enough for us to make changes and sacrifices, so that we do not live at the expense of future generations. Our ability to come together to stop or limit damage to the world's environment will be perhaps the greatest test of how far we can act as a world community. No-one should underestimate the imagination that will be required, nor the scientific effort, nor the unprecedented co-operation we shall have to show.*
> *We shall need statesmanship of a rare order.'*

So, for *twenty-seven years*, both political leaders and the general public have known that the Earth's atmosphere has been undergoing what is effectively a chemical change. We have also known that this change is a direct result of using

---

[50] With permission of the Margaret Thatcher Foundation.
Source: www.margaretthatcher.org.
Reference: http://www.margarettthatcher.org/document/107346.

inexpensive, convenient, CO_2 emitting energy sources to power the world economy.

People in the scientific and political communities have issued frequent warnings of the peril facing humanity if the reliance on CO_2 emitting energy sources continues unchecked. This is akin to knowing the health risks of tobacco and asbestos, but allowing both to go unchecked for many years.

The scientific community has been effective in getting the concept of climate change across to all sections of the population. However, science has not been as effective in conveying to the world leaders and the community, the need to follow up the warnings with action.

The Framework

World leaders and government officials attend high-level international conferences at least annually and often more frequently—some relate to global warming and many do not. Whether the agenda of a meeting relates to economics, trade, agriculture, geopolitical matters or is specifically targeted to climate change, makes no difference. The challenges of global warming and climate change cross all national and economic boundaries.

The important point is that high-level government-to-government conferences also have a high media and public profile—which is essential to galvanise public opinion and obtain public approval for remedial action to mitigate the warming trend.

The role of public opinion is critical in getting action on global warming and climate change. To this end, it is not sufficient to have the world leadership team meeting yearly, or even bi-annually; it should be continuous, similar to high-level war cabinets and general staff meetings in times of world war. Global warming is as grave as a world war and work should not stop until the problem of excess greenhouse gases is overcome. Infrequent and sporadic effort is not good enough.

Starting in 2008, the unfolding financial crisis that followed should provide much food for thought and be considered a warning. If an atmospheric crisis of a similar or greater magnitude to the financial crisis occurs, we may not have the capacity, or the time, to take corrective action.

It is sensible for us to do everything possible to avoid a crisis in the biosphere, just as it would have been sensible for policy makers to have heeded the warnings of a looming financial crisis prior to 2008. Then, as now in the climate debate, those warning that all was not well in the global financial system were drowned out by powerful self-interest groups, along with our natural leaning towards believing that everything would indeed be 'fine'.

The emerging atmospheric crisis is likely to be orders of magnitude more serious than any financial crisis. It is putting a lot of faith in those, who never saw the financial crisis coming, to assume that the same group will anticipate and deal with the economic crisis that will be a consequence of us getting the economics of the biosphere wrong.

Timely and significant action is necessary to avoid progressively more dangerous climate change. This action needs to be on a scale sufficient to counter well over 2,000 billion tonnes of extra CO_2 that has already been released into the atmosphere, during the last two and half centuries of extensive industrialisation. Of course, the emissions of the first century of the Industrial Era were inconsequential compared to the exponential growth of emissions over the past 100 years.

The lack of political determination to implement action on a scale necessary to effectively undo the existing damage to the biosphere is at the very heart of the atmospheric issue.

Science no match for Politics

The scientific method is to observe a phenomenon and then formulate a hypothesis to explain that phenomenon. In a very simplified description of the process, any hypotheses are either supported or disproved by data obtained from observation and experimentation. Results need to be predictable and repeatable and scientific reports must be backed by rigorous proof.

Characteristically therefore, science is limited in its ability to speculate on unlikely, but still possible, scenarios. The nature of the scientific method makes it difficult for scientists to predict and prove conclusively the outcome of a process where input and output parameters are largely unknown. In the instance of the emerging crisis in the atmosphere, one of the significant unknowns is a multiplying and perhaps additive, exponential feedback effect.

The political process, which is not constrained by such self-imposed rigor, appears to demand of science that every part of the incredibly large climate system is known in perfect detail. It seems to be taken as a prima facie case that without such detail, we cannot commence remedial action. This ignores the fact that current science is already sufficiently rigorous for us to demand much more from the economic/political side of the discussion.

> *There is an inequity in any debate between 'science' and 'politics'. The 'political' can speak in unlimited clichés, but 'science' has to prove every last small detail—even when the topic being discussed is a biosphere undergoing changes not seen for hundreds of thousands of years.*

Of course it is appropriate to expect rigor, thoroughness and diligence from our scientists, but the same rigor is not always demanded of everyone in society.

> *Because the scientific community always specify the degree of uncertainty in the supporting data, the only thing the sceptic and the people who deny anthropogenic global warming, need to say is: 'even the scientists are unsure'.*

It seems that the level of analysis and diligence that people demand from scientists about climate change is rarely demanded from politicians in respect of economic and/or social outcomes arising from reports and committee recommendations in the areas of education, health and welfare reform.

Many of these reports and recommendations also have a great bearing on people's lives. In the special case of climate policy decisions, it is recommendations from science that may well determine our survival. If the unequal contest between science and politics is allowed to continue, it will hurt us all.

A very real concern is that as far as we know, we are the first living beings to have developed the capability to significantly alter the environment on which our existence depends. Currently, we have not developed the necessary political processes to manage this immense power.

Lowest Common Denominator Economics

The political process tends to distill debate into short, often emotive media grabs. In the context of the Australian climate debate, frequent use is made of expressions such as 'only by adopting (a particular) policy, will a certain environmental icon, such as the Great Barrier Reef, be saved.'

Simplistic statements are unlikely to foster the type of discussion needed on, for example, the relative effectiveness of different emission trading systems. It is exceedingly unlikely that anyone in the world believes that any action by a single country, acting alone, will make any difference at all to the environmental outcome for any particular national environmental icon. Much more will be needed. This is explored further in later chapters.

Political leaders around the world must now earnestly engage with the scientific position on the climate and *articulate this to the electorate*. If governments and oppositions continue with emotive and simplistic appeals for populist support, there is scant hope for effective action. Those on the denialist side of the climate change debate exploit any 'weakness' in political determination and use such weakness with great effect to challenge effective action. The lack of concrete outcomes from recent climate change conferences supports this contention.

In matters of the science and politics of climate change, advocates for what are effectively, 'business as usual' scenarios eschew important scientific facts by the use of sweeping, broad-brush, but flawed economic assertions. From a global warming perspective, these simplistic statements have been successful in changing the debate about the atmospheric crisis into one about economic well-being and tax. When the scientific debate is redefined as a public policy debate on whether or not people are willing to pay more in taxation, fuel, energy, food and water costs, then the battle to adequately reduce emissions is lost.

It is remarkable how the climate debate has been transformed from the consideration of how we should respond to the possibility of an environmental disaster, into what is effectively an economic debate. Moreover, the economic argument is not even focused on how to fund the extensive scientific and engineering research that is required to avoid unknown and potentially catastrophic environmental outcomes.

Everyone, particularly those who influence or shape policy decisions, should be mindful of the words of the economist, Emeritus Professor Herman Daly, of the University of Maryland:

The economy is a wholly owned subsidiary of the environment, not the reverse[51].'

---

[51] From *Steady State Economics*, by Herman E. Daly. Copyright © 1991 Herman E. Daly. Reproduced by permission of Island Press, Washington, D.C.

'Can Do' Economics

When it comes to global warming and climate change mitigation, there is a discernible move from 'can do', to 'can't do' economic attitudes. This is evidenced by the lack of any meaningful and binding international carbon emission reduction agreements. The change from 'can do', to 'can't do', is hard to understand. It has no logical basis and is inconsistent with the philosophies of the modern world and also most fields of human endeavour.

Currently economic arguments are being used with great effect, to avoid pre-emptive actions for minimising dangerous climate change. It is indeed fortunate that similar arguments did not dominate economic thinking in the past, for example the period 1939 to 1945.

Effective economic engineering is vital for the survival of our civilization. Allocation of both capital and revenue in the future will increasingly need to be determined according to scientific and engineering criteria. *Business as usual* and *solutions as usual*[52] economic models are inadequate to ensure avoidance of a climatic tipping point in the future.

The negative side of the climate debate frequently brings up the rather pathetic catch-cry of how much the mitigation of the crisis in the atmosphere is going to cost. The answer to this negative lobby should be loud and unequivocal: 'Of course saving our civilization will cost money! It may end up being the largest research and development investment in recorded history—and well worth it.'

The economic benefits will follow. First the investment must come.

---

52 The term *'business as usual'* has been widely used for many years and the meaning is obvious. In the context of this book the term *'solutions as usual'* is used to describe proposed global warming solutions that do not run counter to the *'business as usual'* economic model.

Every scientist, engineer and politician, indeed, anyone cognisant of the gravity of the atmospheric situation, has a responsibility to increase awareness within the broader world community, that current science, engineering and dependence on fossil fuel *cannot provide environmentally sustainable answers* to the burgeoning global demand for energy, food and water.

The antithesis between science and politics was well described by a good friend.[53]

> *'Yes, the politicians may say they understand the seriousness of climate change, but they do not believe it.'*

---

[53] With permission: Edward (Ted) Holmes. Management Advisor, Academic and Poet - 2009

7. High Stakes — Myths and Uncertainty

Speculation on what could go wrong with the climate is fraught. It would be dangerous to assume that our current analysis of the atmospheric equation is one hundred per cent accurate in every detail. There is always going to be some uncertainty in the study of a scientific field that deals with phenomena not previously encountered, or at least not encountered during the period of modern scientific analysis.

A consequence of this is that global warming denialists leverage off this uncertainty and gain traction with their assertions that it's nothing we humans are doing that is causing warming of the planet, or even that it is not actually warming. However, based on the weight of current scientific knowledge, the only prudent course of action from a risk management perspective[54], is to take the view that science is giving us a realistic appraisal of the situation in regard to human-induced climate change.

A climate change denialist will frequently speculate that everything will be fine, even if we do virtually nothing. But what if they are wrong? It is more important to consider what is likely to happen if we ignore the very clear warning signs of science and the environment.

Action to negate the emotive and unscientific arguments advanced by those who deny human involvement in climate change is vital.

The most important mission in the world today is to reframe the political and economic agenda, from business as usual, to one that is environmentally friendly.
The world cannot afford to get this wrong.

---

[54] Chapter 17: A Risk Management Perspective.

Habitat Change

Recent research on the common Brown Butterfly in Victoria, Australia, shows the butterfly is emerging from the larvae ten days earlier than was the case for the same species fifty years ago. Dr Michael Kearney and Professor David Karoly attribute this to the temperature rise in this region of Victoria, of about one degree Celsius over the last fifty years.[55]

This research shows that for a butterfly, a change in its habitat will cause a change in behaviour. It is highly likely that these resultant changes will have significant consequences for either the butterfly, and/or some associated species.

Butterflies are not the only organisms with a changing habitat.

Humans are in a more fortunate situation than butterflies. Up to a certain level of change, humans can modify their environment to stay comfortable. However, the very act of modifying our habitat can exacerbate the rate of externally imposed change. Our ability to adapt to changing conditions on the surface of the planet will reach a limit which cannot be safely exceeded.

If the community continues to be lulled into inaction by a perspective that denies human-induced warming of the planet and this position is proven wrong, there will be no going back.

How tragic it will be if at some time in the future we lament: 'But, how were we supposed to know things would get this bad? If only we could return atmospheric CO_2 to 280 ppm, as it was at the start of the Industrial Era, or even the 2015 level of just over 400 ppm?' By then it will be far too late.

---

[55] Michael Kearney, Natalie J Briscoe, David Karoly, et al, Early Emergence in a butterfly causally linked to anthropogenic warming. The study was published in *Biology Letters*, a journal of the Royal Society, 17th March 2010.

Myths and Modern Folklore

Popular myths regarding climate change do nothing to encourage prudent courses of action. One such myth promotes an idea that the climate variation presently being experienced is simply part of a natural cycle and therefore not due to any human activities.

We frequently hear this myth, which says that the climate has been changing for hundreds, thousands or millions of years and that 'we' should just get over it and curb our anxiety. A refinement of this viewpoint suggests that as climate variation has always been a part of the earth's biosphere, not only should we put aside our concerns, but there is nothing we can do about it.

Apart from relatively minor and geographically localised climate variation in the Middle Ages, known as the Medieval Warm Period[56], the disturbing feature of this line of argument is that supporters can only expand on this theme by talking about climate variations dating back to pre-history. For example, a denialist of human-induced global warming will frequently say that the level of atmospheric CO_2 has been higher at previous times in the Earth's past, even referencing situations that date back millions of years.

What possible relevance could millions of years ago—long before the ascendancy of modern humans—have to the development and maintenance of modern society?

It is difficult to be comforted by examples of drastically different climate and ocean acidity regimes that pre-date our comfortable advanced civilizations. It's the Earth's current benign environment that has supported the development of human society, the benefits of which many of us enjoy every day. It is likely that some of the past climate and ocean acidity

---

[56] The Medieval Warm Period occurred about 1000 years ago. This was not a worldwide phenomenon with some regions warmer and others actually cooler. This was a period of increased solar radiation and reduced volcanic activity. The Medieval Warm Period does not compare to the worldwide warming of the last two decades.

levels would have adverse consequences for our current environment and world.

Another dangerous myth is that because only about 3½% of the total emissions, of over 810 billion tonnes of CO_2 per year, originate from human activities, these human emissions do not matter.

It is correct to say that nearly 97% of total emissions are due to natural sources. However, these emissions have been balanced for at least half a million years by natural absorptions. Humans are fortunate that so far approximately half of our extra emissions have been absorbed by an increased natural uptake, particularly by the oceans; otherwise the situation would be far worse.

It is of concern that some people still maintain that human emissions are too insignificant to matter. While it is true that human CO_2 and other greenhouse gas emissions are small in comparison to natural emissions, they are *large enough to upset the natural balance*. They are the proverbial straw that breaks the camel's back. Of even greater concern is that the false cry of those who deny anthropogenic warming still appears to be taken seriously by a number of influential people and politicians.

Some have even argued that higher levels of atmospheric CO_2 will promote faster plant growth and will help solve possible food shortages, as the world population is anticipated to reach nine billion by mid-century.

To argue that the world would be a better place with an atmospheric CO_2 level of 450 parts per million or even more, because of faster plant growth is highly questionable. Even if there is some small benefit in faster plant growth, it will be offset many times over by the negative consequences of a warming planet.

Natural Cycles and Questionable Logic

An effective ploy of a person, who denies human involvement in climate change, is to use a clear and obvious fact in one instance to support an erroneous assertion in a completely different context[57]. For example, a denialist will use the fact that there are many naturally occurring cycles affecting life on Earth, to extrapolate that all atmospheric change arises from these cycle variations. By this simple ruse, the existence of naturally occurring climate events is used to deny the science that clearly shows the effect of some modern industrial practices on the environment.

This is a fundamental flaw in the climate change denial argument and must be countered at every opportunity.

There are many natural cycles. Cycles in sunspot activity, the Earth's orbit and axis of rotation, the Lunar orbit relative to the Earth, air-stream patterns in the stratosphere, high rainfall, low rainfall, drought and maybe even in volcanic activity and earthquake clusters.

These cycles are independently identifiable and it is undeniable that some, many, or maybe all of them have some influence on global temperature and climate. They are long established, relatively constant and have been present during most, if not all, of human development. The 'natural' world is not the problem. The 'natural' world has not caused an increase in atmospheric CO_2 of 43% during the last two hundred and fifty years.

The current changes in climate arise from the exponentially increasing human and industrial CO_2 emissions and the finite ability of the biosphere to absorb these extra emissions. Of course, the situation is compounded when the effect of anthropogenic CO_2 emissions are superimposed on natural planet-wide warming and cooling cycles.

---

[57] The classic non sequitur.

It is always possible that a naturally occurring phenomenon such as a super volcano or errant asteroid may one day have an adverse effect on human civilization. Let us not increase the list of potential dangers that could face humanity, by continuing to add CO_2 to the atmosphere.

It is also possible that a significant cooling cycle (Ice Age) could help to slow the warming trend. However, the onset of a new Ice Age cannot be relied on as a global warming mitigation tool and would hardly be conducive to the continued comfort of the human race.

Compounding the dubious logic of the denialist argument is the rejection of overwhelming and peer-reviewed scientific evidence. Evidence that supports the case that CO_2 emissions, from human industrial and agricultural activity, are causing the rate of CO_2 accumulation in the atmosphere to accelerate and is also causing the observed changes in the climate. It is important to note the lack of genuine peer reviewed scientific research, backed up by appropriate data, which supports the denialist side of the global warming debate.

An integral part of science is a healthy scepticism and the peer-review system is how the science of any topic is refined and validated. This is achieved by scientists working in the discipline independently repeating and confirming the hypothesis. It is astonishing how easily those who deny anthropogenic warming, can ignore the mass of overwhelming scientific evidence that supports a clear warming trend and which also correlates this trend with increasing levels of CO_2 in the atmosphere.

By ignoring well-defined scientific evidence, the role that anthropogenic CO_2 emissions play in global warming is discounted by the climate change denialist. It is concerning that these strategies seem to be working for their argument and a substantial part of mainstream political and public opinion appears to be steadily embracing this dangerous position. As a consequence, this thinking has made it easier for special interest lobby groups, promoting the fossil fuel industry, to weaken any proposed action for effective CO_2 mitigation action at the international level.

Risky Business and Harsh Consequences

Before concluding the 'natural cycles' issue, it must be acknowledged that humans are an integral part of the natural world. It has been argued, that because we are a natural part of the biosphere, then any man-made industrial emissions are likewise a natural part of the biosphere. Following this particular argument to its conclusion is very risky.

If we assume that as part of the natural world, whatever we do is environmentally acceptable, the consequence may be that human beings will have to accept the type of harsh penalty that nature dispenses to other species that overpopulate and become excessive users of crucial, but scarce resources.

An example would be the periodic explosions and collapses in the population of certain species of small rodents, during alternating periods of favorable and unfavorable breeding conditions. A specific example of this would be the recurring plagues of mice in the southern and eastern grain belts of Australia.

Another unfortunate absurdity with the: 'It is all due to natural cycles' argument, is a failure to give due weight to the critical importance of these same cycles. Over millions of years, these cycles and other ambient conditions have been a key in the evolution of the Earth's biosphere and all living organisms, including ourselves. What we are now doing is imposing 'unnatural' external factors on the finely balanced evolved biosphere.

Of course, it would be superb if the denialist argument was correct—that we really do not have anything to be concerned about. However, the continuing and accelerating loss of ice mass from the Polar Regions and the increasing acidification of the oceans provide the reality check that gives lie to this beguiling, but flawed hope. The science is not wrong and the reality of the situation must be urgently addressed. We cannot afford to get the response to such a clear scientific message wrong.

It is time for change.

Crossroads

A good deal of comment has been directed at the use of the word 'change' over the last few years, suggesting it has been over used. Actually it is not so much that the word has been over used, but more that it has been under actioned. Key aspects of change are difference, alteration or modification.

The problem is that *there is a mismatch between the reality and the rhetoric of policy decisions;* this is particularly true in the case of action to counter global warming. While the stated intention is to achieve change, in fact, the constant repetition of the word 'change' is effectively being used as a substitute for action. It is almost as if, by talking about change, a public figure can give the impression that change is occurring and something meaningful is actually being done.

Humanity is at a crossroad where the level of change required is daunting, especially when taken in the context of our present thinking and conventions. In spite of this, at this critical point in civilization's history, the world needs to embrace the concept of planned and programmed change, rather than, through inaction, allowing external forces to ultimately trigger chaotic change.

Chaotic change will follow some unknown environmental aberration in the future and will likely be at a level that is very harmful to our civilization. The idea of planned, programmed and incremental change is explored in later chapters.

Of course, the required rate of change would have been much less if the world had responded to the challenge twenty, or even ten years ago. The cost of dismissing the power of harmful exponential variation is immense. But we cannot travel back in time. It will however be disastrous if we let another ten years elapse before we make substantial alteration to the economic, manufacturing, energy and energy distribution paradigms.

The required level of change to the conventional economics of manufacture and distribution, while daunting to some, will be embraced by many others who have a deep disquiet about how dire the emerging situation in the atmosphere/biosphere is becoming.

So the key question remains: When is something effective going to be done to stop or at least slow, global warming? This is something we cannot afford to get wrong just because of our natural resistance to change. The human reluctance to embrace change is exploited with great effectiveness by the climate change denialist. This unwillingness is especially strong when the changes required are to long-established and comfortable routines.

Easter Island Revisited

There are many indications to suggest we have already got the atmospheric balance wrong and that time is long past for time-wasting debates. It is more relevant to ponder the question of whether we are already setting up a contemporary version of the depopulation of Easter Island, as discussed in Chapter 2.

Despite of all the high-level scientific evidence, as a society, we still strive to deny the danger presented by the warming effect of exponentially increasing the level of CO_2 in the atmosphere.

As a seven billion strong tribe, we do not seem able to grasp the concept that we may be doing exactly the same as the tribe of Easter Islanders did all those centuries ago. The difference now is we are doing it with a vastly superior knowledge of the complex and interconnected environmental systems, with which we are interfering.

Negative outcomes for the biosphere in our time will not be able to be dismissed as a 'how in the world could we have predicted that event'. Serious negative outcomes for the Earth's biosphere are predictable and will result from us altering the chemistry of the atmosphere.

It is possible for us all to research and refine our view of what happened on Easter Island. However, it is not the fine detail of the Islanders' final years that matter, but the underlying lesson of those years. The environment can be a harsh administrator of evolutionary and environmental justice and it is vital that we, in the 21st century, re-learn that lesson before it is too late.

Modelling the Science

How can science 'model that which cannot be modelled'?

As discussed earlier, science is limited in its ability to be specific about the unforeseen. The rigours imposed by the scientific method make it impossible to obtain the experimental data that would be necessary to identify a trend towards what may be a 'once only' and totally unexpected event.

In the closing stages of the 20th century there was the 'Y2K' problem. The identified concern was the date fields on critical computers and the fear was that the changeover to 01/01/2000 could cause computers to reset to a date that was nonsense to the particular software or control system and, as a consequence, could cause great chaos. Enormous amounts of money were spent on Information Technology (IT) upgrades in the years leading up to that very special New Year's Eve. Eventually the dawn of the 21st century came and was largely uneventful from an IT standpoint.

There were vigorous debates, both before and after 00.00.01am on 01/01/2000, as to whether there was actually a real problem, or whether it was mostly hyperbole. While the Y2K debate is not directly relevant to the climate debate, what is pertinent to the current situation is that the IT industry was effectively being asked to give assurances that nothing would go critically wrong, when the numerals in a two digit date field went from 99 to 00.

The difficulty then and today is that professionals in any field of endeavour cannot give an assurance that things will not go wrong unless they can test the problem. So, if the problem cannot be tested, then assurances cannot be given that things will not go wrong. Just as it was impossible to test a change of date from one century to the next, it is even more difficult to test for an increase in CO_2 to, for instance, 450 ppm in the atmosphere of a 5.98×10^{24} kg planet.

Whether or not the reason there was a trouble-free transition from the 31st December 1999 to the 1st January 2000, was due to the vast expenditure on information technology infrastructure, is only part of the point.

The real issue was that businesses and governments, around the world, were not prepared to take the risk of catastrophic computer and information system failure. The question that must be asked now, is 'why do global businesses and world governments, appear to be taking a much bigger risk with the possible catastrophic consequences of global warming?'

Getting Started

Science cannot *definitively* model the unknown and governments *definitely will not* model the unknown. This is because there is a probability of incurring costly scientific and engineering research, or the need to introduce policies that may be unpopular with the electorate. Crazy though it seems, the chance of an end to civilization at some date in the future, is more palatable to a modern government than the prospect of losing office at the next election.

The result is that any change to public policy, regarding CO_2 emissions into the atmosphere, is going to come down to the average people that we meet every day, all over the world. Moreover, it will only then occur if we make our protest with a loud, clear and cohesive voice.

The case for strong action needs to be taken into all of the places we work and live because the political debate will follow the 'people debate'. In the strange new world of current politics, it seems to be impossible for the politicians to lead any debate that involves increased expenditure on civilian infrastructure.

Everyone in the world who is concerned about the warming globe should be unapologetic about repeating the message of just how dangerous global warming is to our society.

Repetition—for years, so very effectively, this is exactly the tactic of those on the denial side of the anthropogenic global warming debate.

In the banks and in the factories, in the homes and in the schools, in the streets and in the shops *and in the lack of compelling, determined political action,* the denialist side of the debate still appears to be winning.

8. Melting Ice

An important 'engine' of world climate is the great mass of ice, 30×10^{18} kg, stored in the polar ice caps and ice sheets[58]. As the planet warms, the ice in the Polar Regions will melt, including the land-based ice sheets, which in turn will lead to rising sea levels. This is a critical issue, because it is the melting of land ice, rather than sea ice, that does the damage to shorelines[59].

The majority of the world's land-based ice is situated in the Greenland and Antarctic ice sheets and although they may be melting at different rates, they are in fact both melting— this is a fact. As these two massive land-based stores of frozen fresh water melt, sea levels will inevitably rise, inundating much-needed coastal agricultural land and displacing populations.

The rising world population will exacerbate the effect of loss of costal land suitable for industry, commerce, housing and agriculture. Any land given up to the sea will be sorely missed. The implications of rising sea levels will be further explored in Chapter: Food, Water, Refugees and War.

Faster and Faster

Information about the rate at which the above two great ice sheets are melting is readily available in the public domain and it consistently points to the ice melting at an ever increasing rate.

Those who deny the evidence of a warming planet have made much of some increased snowfall, over parts of Greenland and East Antarctica. However it is not remotely

---

[58] Including Greenland.

[59] Sea ice is already floating in the sea and consequently will not cause in increase in sea levels any more than a melting ice cube in a drink will cause an increase in the drink level in the glass.

possible that this will compensate, in any meaningful way, for the loss of the massive land-based ice sheets. Undeniably, the world has lost thousands of billions of tonnes of ice from both the Arctic and the Antarctic, over the past couple of decades.

There are varying estimates of the time span until the great ice sheet of Greenland will disappear. It is worth noting that there are scientists who believe that slippage of land-based ice sheets may be accelerated by melt water permeating the ice sheet and making the interface between the ice sheets and the underlying rock more slippery[60].

Further, there are many estimates of when—not if—the floating sea-ice in the Arctic will progress to rapid summer disintegration. These estimates vary from within a couple of years, to two or three decades.

One thing is sure; as the polar ice cap melts, more of the Arctic Ocean will be exposed to solar radiation and even if complete summer disintegration of Arctic sea-ice doesn't occur exactly in a particular year, partial disintegration is almost certain. The rapid reduction in volume (thinning) of the Arctic ice is of great concern and makes the triggering of an 'ice-melt' tipping point more likely than not.

With the rate of ice melt increasing exponentially due to temperature alone and the risk of water assisted slippage of glaciers and ice shelves further speeding things up, we need to move much faster to correct the situation. As actual climate and temperature data is updated, many scientific predictions about the changes in the physical characteristics and chemistry of the Earth, due to global warming, are more worrying than originally predicted. This applies in particular to the Polar Regions.

The role of the Arctic as an 'engine' of Northern Hemisphere weather is another important consideration that should not be ignored.

---

[60] This applies in both the north and south, Polar Regions.

There are other changes to the biosphere that may be accelerated by an increase in average global temperature. For example, the progressive melting of large areas of permafrost in the high latitudes of the northern hemisphere will rapidly release the other major greenhouse gas, methane, from stored frozen decomposed vegetation.

Warming of the oceans may also cause methane to be released from certain ocean floor locations in the Arctic and the Gulf of Mexico. There is the possibility of a catastrophic methane driven tipping point occurring.

Albedo

Yet another effect of rising global temperatures will be a reduction in the average Albedo value of the Earth[61]. As the albedo is reduced, less of the Sun's heat is radiated back into space which causes further warming of the planet. If the earth were a perfect reflector, the albedo would equal one[62]. The Earth is not even close to being a perfect reflector, which is fortunate for us because if it were, we would not be here. With an albedo of one (unity), all of the Sun's heat would be reflected back into space leaving the Earth as a frozen planet and not the beautiful blue/green oasis in space that we love so much.

The albedo of planet Earth varies with different surface features. Albedo values vary from zero to one and the higher the value, the more solar radiation that is reflected back into space—the lower the albedo, the more that solar radiation will be retained and will heat the planet.

The overall planetary average is somewhere between 0.3 and 0.35 or stated differently, about a third of the Sun's energy is reflected back into space. Clouds and snow have the highest albedo values; up to 0.7 for dense clouds and up to

---

[61] Albedo is a measure of the proportion of the Sun's heat that is reflected back into space.

[62] For reference, the albedo of a perfect 'blackbody' would equal zero.

0.9 for fresh snow. Albedo values of around 0.1 to 0.2 are typical for forests, soil and crops whereas the albedo value of deserts is around 0.3.

The albedo of sea ice varies depending on whether it is covered with fresh snow or not, but without snow, the average is similar to dense clouds. The albedo value of the open ocean is approximately 0.05 (only 5% of solar radiation is reflected back into space).

The 10:1 albedo ratio[63] of sea ice to open ocean is problematic. If large areas of sea ice continue to melt, the average albedo for the planet will be reduced, resulting in less heat being reflected back into space and further contributing to a warmer planet.

Another factor that influences the albedo value is the angle of incidence at which sunlight reaches a particular area of the Earth and whether it is land, a turbulent ocean or a calm ocean. The albedo balance is far from simple and the fact that deserts have a higher albedo value than forests, does not suggest that creating more deserts on our planet is a solution to global warming. Apart from the role they play as CO_2 sinks, forests also contribute to other complex feedback loops. Any change to the environment of the world's forests may trigger detrimental environmental tipping points.

Considering the Improbable

Popular media has a major role in the moulding of public opinion and hence the formation of public policy. It is through the media that most of us get our information; very few people read scientific papers and abstracts. Unfortunately, any scenario that might be catastrophic for civilization has been relegated, by popular media, to the sphere of entertainment. There are many doomsday scenarios in books

---

[63] This is a minimum ratio - if sea ice is covered by fresh snow the ratio will be higher.

and movies. Usually something miraculous happens at the eleventh hour to save the day.

Treating disaster scenarios as entertainment devalues the currency of predictive analysis. Apart from some media articles exploring and maybe even sensationalising the possibility of an errant super-bug or asteroid taking us out, there is little interest in this area of analysis.

It is risky for anyone to speculate on improbable, but potentially catastrophic events, for fear of being labelled crazy, an alarmist, a doomsayer or sensationalist. This is troubling, as in addition to the usual science on global warming, there is the possibility that some improbable, but extremely damaging consequence may result from anthropogenic changes to the atmosphere.

Improbability alone is no guarantee that such an event will never occur. The only thing improbability does is that it limits mainstream scientific and political debate.

One highly speculative example of an improbable event is the possibility that redistribution of the planet's ice mass, as a result of ice melt, could cause a change to the wobble in the Earth's axis. This possibility ought to justify at least some research into the stability of the Earth's axis.

It is known that the axis of the Earth wobbles and that there are also a number of superimposed wobbles with repeating cycles and varying time periods. There are also other factors that influence the wobble of the axis. For example, it is reported that some hurricanes and earthquakes[64] have been large enough to have a measurable effect on the 'wobble' of the Earth's axis.

The mass of the Greenland ice sheet is approximately 2.7×10^{18} kg and while this represents slightly less than a two millionth part of the Earth's total mass, it is still a very large concentrated mass. It seems reasonable to assume that such a mass, displaced approximately 25° of latitude to the axis of

---

[64] For example: the Christchurch, New Zealand earthquake in 2010.

rotation could be significant in any calculation concerning the rotational mechanics of the planet.

The effect of melting away a 2.7 x 10^{18} kg mass of ice is effectively the same as adding exactly the same mass at the same latitude—we would most likely be hesitant about doing that! The calculations required to predict axis behaviour, taking into account the possibility of total Greenland ice loss by the end of the century, or earlier would be exceedingly complex. Maybe nothing can ever happen to the axis of a 5.98 x 10^{24} kg planet, but the question still remains: 'Can we be assured that partial or total loss of the Greenland ice sheet will not have a destabilising effect on the axis of the Earth?'

9. Science and Engineering — Some Background

Setting the Scene

We exist in two parallel worlds, the natural world and the manufactured world. Co-existence between the two worlds is an increasingly thorny issue. Even what might appear to be 'natural', such as a field of wheat or quietly grazing cattle, owe much to the manufactured world such as harvesters, antibiotics, grain silos and tractors to name a few. The dawn of the 'manufactured' world, in its simplest form, pre-dates recorded history. For millennia, the influence of the manufactured world on the natural world was negligible, but this is no longer the case.

Almost everything we see and use is a product of manufacture and would not exist if not for science and engineering. We are where we are today, because science and engineering enable our comfortable lives.

The developed world has wholeheartedly embraced the benefits of science and engineering. Increases in atmospheric greenhouse gases are a consequence of this. Ultimately the stability of society will depend on science and engineering, together with our current and future resource management strategies and priorities.[65].

Part of the climate change debate should include how we manage the growing energy and material resource requirements that are associated with rapidly increasing consumer demand for manufactured goods and services that have a relatively short serviceable life.

---

[65] In this context 'resources' includes the resources that are consumed by our 'industrial society' and also the 'natural' resources of the planet that are essential to maintain the biosphere in a condition suitable for human habitation.

Tackling the Problem

Global warming has arrived and until now, has been almost symptomless in terms of the effect on the industrialised world. The symptoms however, have been getting more and more recognisable on the natural world around the globe and because the changes are planet-wide, the reversal of the warming trend is almost an intractable problem. Because of this, there is good reason to believe that no single 'yet to be discovered' scientific breakthrough nor existing technologies, *taken separately*, will do the job of preserving our benign biosphere.

A wide range of strategies, *taken together*, must be directed to the job of reducing atmospheric CO_2. This includes multiple future scientific and engineering developments, as well as all of our existing CO_2 mitigating technologies.

Every possible strategy must be directed to the job of maintaining the human-friendly biosphere.

Until new zero emission base-load energy technologies are discovered, developed and fully deployed on a planet-wide scale, we must fully utilise the known low/zero CO_2 emission energy sources. Energy conservation measures must be used to the fullest extent.

Wind, Water, Rock and Sun

Currently, the major energy sources that are low in CO_2 emissions include wind power, hydroelectricity, nuclear, geothermal and tidal as well as various methods of utilising the radiation of the sun[66].

Because of the need for supporting infrastructure, there are no sources of renewable energy that can be used on an industrial scale that do not entail at least some level of CO_2

---

[66] Solar energy sources include photovoltaic cells, direct concentrated thermal solar, solar thermal/molten storage etc.

emissions. This is an important consideration in the selection of low emission and renewable energy sources.

The construction of nuclear power plants is a particularly CO_2 emission intensive activity. Manufacture of wind turbines and construction and siting of the towers, especially at sea, are certainly not without a CO_2 emission cost. Mining and/or refining materials used in the manufacture of photovoltaic cells or sheeting for solar power, also have an emission cost and this is before the capital CO_2 emission cost of the factories, the manufacture and transport of the finished product is considered.

It has even been said that in south-eastern Australia, during a particularly long dry spell, electricity generated by some very high CO_2 emitting brown coal-fired power stations was used to pump water back up the mountainside, for use in a hydro-electricity power station storage dam. While such practices may be expedient for short-term emergency measures, they clearly do not have a lasting place in a CO_2 emission constrained world.

It is crucial that the global warming mitigation task is undertaken in an environmentally sustainable way. The least favourable outcome would be to arrive at a solution to one environmental problem, only to create another.

In addition, methods of storing energy during periods of low energy demand are technologies that need a good deal more research. This is particularly desirable if the sources of energy used to power the off-peak energy storage are in themselves low in CO_2 emissions. Research into energy storage, specifically solar, for use during periods of low solar availability, or high-energy demand is essential.

A balance must be found between funding for the implementation of all *current technology renewable energy sources and conservation measures*, and the critically important expenditure on research into the more *advanced renewable energy technologies*, which will be necessary to secure the longer-term future.

Energy and Numbers

Determining the proportion of total world energy that is capable of being met by non-fossil fuel energy sources (from currently available and immediately foreseeable technologies) in 2050 or even in 2030, is at best, an estimate rather than an accurate calculation—there are simply too many unknowns.

There are unknowns on both the supply and demand side of the energy equation. For example, the demand for electrical energy and portable (fuel) energy in developing nations could increase exponentially.

Further, if carbon capture and geo-sequestration storage technology (CCS) become available on a scale that could enable coal and gas power stations to continue to supply ongoing base-load power, this would demand a significant percentage of the output of the originating power station.

Because CCS is not yet available on a scale that is commercially viable, exact energy requirements are also not available; however, estimates of 30% to 40% of the output of the host power station are frequently mentioned. Current estimates for the proportion of total world energy capable of being met by non-fossil fuel energy vary enormously. For 2050, it is not difficult to find estimates that vary from a low of 15% to a high of 100%.

Assuming a scenario where the economic expansion of the last fifty years continues and also reaches more of the world's population, it is fair to say that the world is looking at a very challenging scenario.

Despite the lack of precision highlighted above, under the current 'business/solutions as usual' economic models, it is clear that investment in the innovation required to develop new methods of harvesting 'non-fossil fuelled', base-load energy, is insufficient.

Whether renewable energy meets 15%, 50%, or 100% of total world energy demand in the years leading up to 2050, depends on multiple factors.

The most critical of these are:

1. the 'cleverness' of the developed and deployed technology

2. actual world population from 2020 to 2050 and beyond

3. total world energy demand—which is inexorably linked to the percentage of the world population enjoying 'first world' living standards, over the next three and a half decades

4. financial and human resources committed to the task.

In terms of Point 1, the 'cleverness' of the technology, given time, renewable energy sources are likely to have the potential to meet most of the world's energy and fuel needs.

Available time is the critical issue in achieving this and currently the speed at which the world is transitioning from fossil fuels to '100% next generation' energy and fuel technology is far too slow. Further, we simply do not yet know the methodologies that will enable enough electricity, portable transport fuels, food and water to be made available to a possible nine billion people, with near zero net anthropogenic CO_2 emissions.

It would be wonderful to think that human society is excited by this challenge. Who knows what the solution will be? It may be large solar thermal storage plants generating electricity in the Gobi, North African, US and Australian desert 'solar hot spots' and distributed worldwide, with negligible losses, using 'superconducting' transmission methods. Or, it could be something as exotic as fusion reactor technology, making use of helium-3 (He-3) as a fuel source, mined on the Moon and shipped back to Earth[67].

No one knows all the answers. But this we do know—we should not be relying on fossil fuels as the principal energy source, for even the next decade. The methods, by which we generate electricity, provide surface transport fuels, produce food and desalinate water in the future, may be just around the corner. We should be prepared that the cost of achieving this will be quite high.

New manufacturing methods will reduce the demand for raw materials by eliminating wastage during manufacturing processes. While excess material that is removed during manufacture is ultimately recycled, it still has to be mined, extracted and shipped. All of these processes are energy intensive, including the process of recycling.

A new manufacturing process that promises to eliminate some of this waste and at the same time reduce costs is *additive manufacturing (3D printing)*.

The next factor, Point 2, deals with the other significant variable in world energy demand calculations. This is the Earth's actual population at different times into the future. The estimated world population in 2050 ranges up to nine billion people.

With the increasing demand for more energy, in particular electricity, the people who manage energy must scope requirements for a population somewhere between the current seven billion and a projected nine billion or more, people.

The total world demand for energy will also depend on the energy requirement of each person on the planet, at any given time in the future. The difficulty is that the average

---

[67] Helium 3 is a potential fusion reactor fuel - an example of the extreme science discussed in Chapter 12: Science and Engineering – The Future. It is very rare on Earth, but it is thought to be more abundant in the surface layers of the Moon.

energy requirement per person (energy intensity[68]), over the coming decades is heavily dependent on the economic advances made by those currently underprivileged in terms of their living standards. This is a significant variable in the 'total world energy demand' equation.

With all these factors, the question of future energy requirements depends on a mixture of politics, economics, science, engineering and population, as well as a multitude of ethical and moral considerations. It is not possible to accurately scope all of these factors from where we are in 2015. It is possible, based on what we do know, to conclude that the only sensible course of action for the human race, right now, is to listen to what reputable climate science is telling us and implement strategies designed to negate any possibility of a worst-case scenario developing.

'Hoping for the best' has never been an intelligent strategy for dealing with an impending crisis and should not get a moment's consideration in connection with global warming— not even as Plan B. The thought that *'hoping for the best'* is ever contemplated as Plan A is very worrying, although the denialist position comes dangerously close. In fact, some who deny a human involvement in global warming do not appear to be even hoping for the best; they just assume there isn't a problem.

Most of us do not apply 'hope for the best' or 'assume the best' on a micro scale in our own lives. We know that the risk of our house being robbed or burning down, or our car being involved in an accident, are unlikely events. But, as individuals and society as a whole, we consider it prudent to insure against these events. And yet, on a macro scale, when we are considering possible calamitous events with our planet, we seem happy to leave it to chance.

---

[68] In economics the usual definition for energy intensity is 'units of energy required to produce one unit of Gross Domestic Product (GDP)'. The GDP, in turn, correlates with the size of a given population and the general level of prosperity within that population.

Without delay, the number one priority, for governments around the world must be the facilitation of the high-level scientific and engineering research necessary to speed up the discovery and deployment of energy and fuel technologies that have effectively a zero net CO_2 emission. Clearly, in the modern world, infrastructure is no longer an exclusively government domain. The massive engineering projects, which the world should now contemplate, will generate many new partnerships. Business, industry, industrial research centres, universities and government, will all enjoy an exceedingly busy future.

10. Science and Engineering — Models and other Information

Over the past twenty-five years a great deal of scientific research, analysis of data and climate modelling has been undertaken. Over that time the steady warming of the planet has been a consistent phenomenon. Numerous factors have played their part in the interaction between the industrial society and the atmosphere and never more so than in the late 20th and early 21st centuries.

For the first half of the 20th century, there was massive air pollution arising from the exponential increase in industrial activity. This pollution consisted of a combination of many different particulates, aerosols and sulphur dioxide (SO_2). During this time, the rate of industrial expansion was unprecedented, fuelled by rapid obsolescence of equipment due to advancing technology. Compounding the industrial expansion beyond that necessitated by obsolete industrial equipment, was the need to arm, equip, supply and fuel two world wars.

This expansion was carried out using very polluting methods of power generation, manufacturing and transportation. During the first half of the 20th century, these extremely dirty and polluting industrial methods may have mitigated slightly the rate at which average global temperatures were rising. The massive particulate and sulphur dioxide pollution generated[69] would have acted to minimally decrease the warming effect of the equally massive emissions of CO_2.

This led to some speculation in the 1970's, that there could have been a global cooling trend sufficient to lead to a new Ice Age. With the benefit of hindsight and greater scientific understanding, we can now see that, in fact, there

---

[69] Sulphur dioxide and particulate pollution, from past major volcanic eruptions, have caused considerably cooler conditions over wide areas of the globe.

was no cooling trend. There was a small slowdown in the *rate of global warming that was interpreted as 'global cooling'.* This very slight slowdown in the rate of warming did not actually cool the planet. Instead the slow, inexorable warming was continuing unabated, even if there was a tiny slowdown for a few years due to the fact that heavy industrial pollution reflects more of the Sun's heat back into space.

The clean-up of 'dirty' smokestacks and other heavily polluting industrial practices, started gaining momentum from the 1960's and we then witnessed in the latter decades of the 20th century and early 21st century, an increase in the rate of global warming once again. Similar reasoning may also help to explain observations that are seemingly incongruent today.

For instance, there is a contention that, from the first decade of this century, the globe is again cooling and this is similarly erroneous. At best, for the denialist argument, the *rate of warming* may be slightly less and even that depends on the geographic location from which the data is gathered and a host of other assumptions.

It is worth reflecting on the fact that much of the world's manufacturing is now carried out in countries that are still in the very early days of environmental clean-up. This new instance of 'dirty' industrial pollution, from the new centres of manufacturing, may once again be slowing, very slightly, the rate at which average global temperatures are rising. This will change yet again as the new manufacturing centres reduce their industrial pollution.

Nevertheless, the current changes in climate are occurring at a far greater rate than can be explained by any natural warming and cooling cycle. If the chemistry of the atmosphere was not being influenced by humans then maybe, in the distant future, when the next major natural cycle occurs that could have the potential to be harmful to life on Earth, we may also have developed technology to a stage where that harm could be mitigated. What is clear is if human beings bring on calamitous climate change in less than 20 or 30 years, *we shall not be ready for the consequences.*

There is a view occasionally espoused, that essentially says that bad things happen and if that's what is in our future, then 'so be it'. Most people though, are unlikely to be so cavalier with humanity's continuing existence. I certainly believe we have a destiny that will be best served by avoiding a premature destabilisation of the Earth's current climatic conditions.

There is also the issue of integrating data obtained by direct observation, with data generated by computer modelling. An excellent example of this is the complexity in reconciling the two sets of information regarding the 'hole' in the ozone layer over the Antarctic in the early 1980's. Ground-based observations and data collected by high altitude, special purpose balloon technology showed that, by the end of southern hemisphere winters, the ozone level decreased markedly.

The decrease was from about 300 Dobson units[70] in the early 1970's, to approximately 200 Dobson units in the early 1980's; a massive reduction for such a short time. However, the scientific modelling at the time, taking into account increasing levels of CO_2, nitrous oxide, methane and the halocarbons, predicted a much more gradual depletion of ozone.

As with the situation faced today with CO_2 and other greenhouse gas emissions, doubt was thrown on the process and the validity of the scientific assertions being made at the time. The inconsistency between the observed data and the scientific modelling was eventually reconciled, with the observed data being proven correct.

The early modelling had not taken into account the fact that polar stratospheric clouds[71] could provide a platform or surface, for the complex chemical reactions that produced

---

[70] The Dobson unit is a measure of the amount of ozone in the atmosphere.

[71] Present in greater quantity in the Polar Regions in the extremely cold winter months.

free radical chlorine molecules that could combine with and destroy the ozone molecule.

This ground-breaking discovery helped the process towards the signing of the Montreal Protocol in 1987. This Protocol banned the industrial use of a wide range of ozone depleting substances[72], within various specified time frames. There was clearly a spirit of cooperation abroad in 1987 to protect the environment that is not being replicated today.

The lessons from the Montreal Protocol, for the banning of ozone depleting substances, are substantial. The contrast between the Montreal Protocol and more recent conferences to obtain agreement to control human industrial carbon dioxide emissions is unmistakeable.

While the drop in ozone layer thickness, from the 1970's to the 1980's was much more rapid, in percentage terms, than the increase in atmospheric greenhouse gases over the Industrial Era, it is important to note that the rate of increase of atmospheric CO_2 escalated as the 20th century progressed, thus heightening the similarity between the two examples.

Large percentage increases or decreases in a chemical constituent of the atmosphere cannot be allowed to continue, without decisive action to remedy the situation. This was recognised by the Montreal Protocol; a shining example of international leadership and cooperation.

It is self-evident that the complete banning of fossil fuel burning would be a considerably more significant economic and technological step, than banning the use of the particular halocarbon group of chemicals. The timeframe for such action would have to be longer than the Montreal Protocol/halocarbon timeframe. So, with currently available technology, a complete immediate ban on fossil fuel use is

---

[72] Including a range of the principal halocarbons such as CFCs, HCFCs and HFCs - chlorofluorocarbons / hydro chlorofluorocarbons / hydro fluorocarbons respectively

obviously out of the question. Consequentially, the actions are:

1. speed up the development of the zero carbon emission technologies that we know will be essential in and for, our collective future

2. greatly speed up the elimination of fossil fuels.

It is likely there will be some exceptions to developing technologies to speedily replace all fossil fuels, for example the aviation industry. Alternative propulsion systems for mass air transport will go to the very edge of any science we can currently imagine. The aviation sector, as a special case, is discussed in Chapter 16: Sustainable Manufacture.

Noctilucent Clouds and Uncertainty

It has been claimed that the mesosphere[73] may be getting colder due to an increase in polar mesospheric clouds—also known as noctilucent clouds. Whether this phenomenon, if proven, would help to combat global warming is unknown at present. Ultimately, it may be determined that noctilucent clouds are not at all pivotal in the global warming equation; except, to highlight once again, that we cannot wait until every last piece of the atmospheric puzzle is in place before starting on the elimination of fossil fuels.

At present, the political agenda for this most important issue is being set by interests outside of the political sphere. Perhaps we need a simpler, more basic approach. When it comes to high-level science, no one can compete with scientists in the field, but what everyone can do is their utmost to help reframe the dominant political discourse regarding global warming.

---

[73] The mesosphere lies approximately 50–80 kilometres above the Earth

The Genie is Escaping the Bottle

Naturally occurring CO_2 has been balanced by *naturally occurring absorptions* for millennia. The human race is now upsetting this long-established balance. It is foolish to assume that the biosphere will adequately negate our emissions forever.

Successive reports in the media suggest that the rate of global warming is always worse than is stated in published reports, because of the time lag in gathering data and building consensus. In particular and as previously discussed, the three principal greenhouse gases have been growing exponentially since the beginning of the twentieth century.

However, the exponential growth in greenhouse gas has not been matched by a corresponding physical rise in average global temperatures. It would appear that complex interactions within the biosphere have been cancelling part of the effect of the added greenhouse gases. The most significant of these interactions is the absorption of a massive part of excess CO_2 by the world's oceans.

The world's oceans may not save us forever. As the oceans gradually warm to greater depths and simultaneously absorb a significant component of anthropogenic CO_2, their ability to maintain stability of temperature and chemistry will be reduced. With every action there is a reaction somewhere in the system. As previously mentioned, the acidification of the world's oceans has also increased by 30%[74] since the start of the Industrial Era, with a measurable effect on ocean chemistry.

The change in equilibrium conditions of calcium carbonate ($CaCO_3$) in the oceans, because of the decline in ocean pH, will have a quantifiable effect on the skeletal and shell structures of marine creatures such as coral that need

---

[74] A pH of 7 is defined as neutral. Increasing acidification of the oceans means that the oceans are becoming less alkaline (less basic). The current ocean pH of approximately 8.1 does not mean that the oceans are actually acidic.

$CaCO_3$ to build those hard structures. The changes to ocean chemistry should not be a surprise; it is a predictable characteristic of non-linear feedback systems.

Furthermore, we are unlocking, at great speed, the vast amount of carbon that was safely locked away (sequestrated) by environmental processes hundreds of thousands, if not millions of years ago. Human evolution is part of these larger evolutionary processes that have occurred in the biosphere.

There is worldwide research seeking to identify ways of re-sequestrating[75] the vast releases of CO_2 that result from the combustion of hydrocarbon fuels, which currently provides most of the base-load power to the modern world. Unfortunately, hydrocarbon fuels are still necessary under existing economic paradigms, as our best efforts to increase renewable energy sources are currently inadequate.

Carbon capture and storage processes (CCS) would consume, according to some estimates, between 30% and 40% of a power station's total output. There will need to be considerably more scientific and engineering research and development, before CCS is close to being a viable option. If renewable energy could be harnessed to power the CCS process it would be a significant breakthrough, but research may be better directed to developing methods of using renewable energy directly for the supply of the world's base-load power.

Science and Technology: New Models/New Future?

The sheer volume of media commentary and books on the subject of global warming may lead a person to consider that we already have adequate scientific and engineering expertise to reduce CO_2 emissions and thus prevent a potential atmospheric crisis. Under this scenario, all we need to do is

---

[75] Various methods of carbon capture and storage (CCS) are the subject of ongoing worldwide research. Despite a few pilot plants the research is a long way from proving full-scale industrial viability.

implement the technologies currently at our disposal, including wind, carbon capture and storage processes (CCS), nuclear, geothermal, tidal and currently available, solar technologies—while simultaneously practicing more rigorous conservation measures.

I challenge the premise that the current economic paradigm and CO_2 mitigation measures will fully counter the problem of the chemical changes occurring in the Earth's atmosphere and oceans. Present science and engineering are not at the level needed.

Relative to the magnitude of the emerging problem in the biosphere, we are taking very small steps. If the principle of the well-known real estate maxim of 'location, location, location', were to be applied to the preservation of the biosphere, it would be 'research, research, research'.

In parallel with original research into state-of-the-art projects to decarbonise the global economy, every existing renewable technology and all possible conservation measures are essential to help maintain climate stability in the shorter term—until new energy and fuel regimes are developed and deployed.

Because of the extraordinary volume of high-level research, on all topics, currently being carried out by scientists around the world, it allows us to assume that all is well in the 'scientific research world' and that any research is well-funded.

However, the percentage of scientific and engineering research being undertaken, that actually has the objective of preserving or returning the biosphere to 'pre-industrial' conditions, is only a very small proportion of the total funding pool.

It is difficult to obtain exact numbers in an area with so many variables and interpretations of what constitutes original renewable energy research, and what is simply researching more efficient methods of using or extracting existing energy sources. I believe the following will give a considered understanding of the spending on actual renewable energy research and development (R&D).

A reasonable estimate of the total annual global spending on R&D is between 1½ and 2½ trillion dollars (US) across all areas of research. The gross amount includes every area of human activity: medical, consumer goods, military, transportation, communication systems, information technology and space travel to name just a few.

Some components of this research reduce the carbon footprint of equipment/devices being researched, but this is a fortunate by-product rather than the principal aim. Conversely for the environment, a proportion of R&D spending notionally designated for renewable energy research, will be targeted to reducing the cost base of existing energy infrastructure. A further part will be targeted to existing renewable energy technologies and not the ground-breaking advanced research needed to eliminate fossil fuels from the global economy.

I doubt that the total R&D spending, devoted to pure original research into technology to completely decarbonise the global economy, is more than ½% to 1% of the total research amount. Even using 1% as a guide, this would equate to no more than US$15 billion to US$25 billion for research specifically targeted to eliminate fossil fuels from the power generation and transport[76] sectors.

Our expectation should be that the elimination of fossil fuels will occur a great many years before 2040. Stabilisation of the chemistry of the biosphere should be the number one priority for everyone. Even US$25 billion represents only about 0.025% of Gross World Product (GWP) of approximately $100 trillion—25/1000ths of one percent of GWP. If this very rough 'ball-park' number[77] is correct, it is

---

[76] The aviation sector, as a special case, is discussed in Chapter 16: Sustainable Manufacture.

[77] It is unlikely to be significantly greater than this amount.

manifestly insufficient to 'decarbonise' the world's energy and fuel cycle in a timely manner.

While high profile, high-level science does attract substantial funding, it does not follow that this funding is targeted to speed up the development and early deployment of the technologies needed to actually eliminate CO_2 emissions.

Deadlines

The need for new high-technology solutions to counteract global warming cannot be overstressed. Right now, it is entirely possible that humans may have less than a quarter of a century to carry out what, in more normal times, may be eighty to one hundred years of science and engineering. This is the challenge that will set the deadlines for the future.

There are no current deadlines for expediting the zero emission technology of the future; there is not even agreement that it needs to happen. There was a deadline on the Manhattan and Apollo projects[78] and those deadlines accelerated completion of these projects.

The first deadline arose from dire concerns regarding national security and the second from matters of national prestige. There should be no doubt that both projects would have taken considerably longer without such tight timelines.

The need to prevent aberrant changes to the Earth's climate ought to accord the highest possible priority to establishing deadlines for the deployment of the appropriate science and engineering.

If the intent of the expenditure is to accelerate the discovery and deployment of technologies that may otherwise

---

[78] The project to develop an atomic bomb in WW2 was codenamed 'Manhattan' and the Apollo project landed a man on the Moon in July 1969.

be expected by the late 21st century, the current level of funding is manifestly inadequate.

Looking Ahead – What More?

Right now, the first priority should be to establish a chain of extraordinarily well funded, 'zero CO_2 emission' research facilities, strategically placed around the globe. The work of these research centres would be to find comprehensive scientific and engineering solutions that are designed to prevent the continued warming of the planet and make the current 'business/solutions as usual' economic paradigms obsolete.

If it becomes apparent that it is impossible to develop new energy and fuel systems quickly enough and despite our best efforts, the biosphere still ends up in a highly compromised state, at least we shall have done everything possible to avert the situation. While this may be small comfort, it will be better than adopting a minimalist approach and being taken by surprise if worst-case scenarios emerge.

Where is the good news?

Unfortunately, there is no good news on global warming yet. The news just keeps getting worse, which is not surprising bearing in mind the phenomenon of non-linear feedback.

In summary, from Chapter 2, the average global temperature has increased by almost 0.9°C, compared with the pre-industrial level. Over the same period, atmospheric CO_2 (now above 400 ppm) has increased by 43%.

Because both the pre-industrial and current CO_2 levels have a base of zero, the 43% is by definition an absolute measurement. By contrast, on the absolute temperature scale (Kelvin), the Industrial Era temperature rise of approximately

0.9°Celsius represents only a little more than a 0.3% rise relative to pre-industrial temperatures[79].

This temperature to CO_2 lag supports the notion that we may be approaching a 'tipping point', should the average global surface temperature respond further to the CO_2 already added to the atmosphere.

We can be reasonably confident that this will not be a linear relationship, making the sensible course of action on anthropogenic CO_2 emissions self-evident. Maybe we should all start to read scientific reports, instead of watching horror movies, to get the heart racing and the adrenaline pumping. At the very least, everyone should take careful note of what is happening in our biosphere.

The world is currently conducting an unintentional experiment on the Earth's atmosphere, without any environmental controls and without contingency plans for shutting down the process if it all goes horribly wrong. With reports of areas of grass now becoming established in Antarctica and the possibility of a year-round navigable Northwest Passage, how many more warnings are needed?

Paying the Piper?

Significant action will be required to restore balance to the environment and this must be led by high level science and engineering. Right now, the human race has a choice about how climate change is managed.

The best scenario is one where the organisation and management of the impending changes are planned and carried out in an ordered manner by ourselves. This will require a corresponding financial outlay.

---

[79] Whenever a percentage change of one temperature to another is calculated it is an absolute scale of temperature that should be used. Using the Kelvin temperature scale, zero equals minus 273.6 degrees Celsius

The alternative is to leave these changes to the natural processes of the planet, in which case the adjustment may well be chaotic and the cost is likely to be prohibitive. Nature does not have the protection of our comfortable consumer society as its foremost priority.

Homo sapiens evolved to suit the biosphere, the biosphere did not evolve to suit us.

11. Don't Hold Back

The task of supplying all of the world's houses, cars, shopping malls, buses and trains with power or fuel is only part of the challenge facing the world today. The real challenge is making, building and when needed, replacing all of these things. This requires immense industrial infrastructure: blast furnaces, petrochemical plants, heavy machinery, thousands of mining and drilling sites, millions of hectares of factories—the list goes on and on.

It is not only about cars, toasters, clothing and computers. It is also super tankers, road-making machinery, aircraft carriers, cement, airplanes, shopping malls, complete cities and even missions to Mars; a seemingly infinite list.

Adding to the complexity of the climate situation, people the world over are acquiring more and more material goods and services. The market is expanding in both numbers and purchasing power. Today's industrial infrastructure will need to expand to meet the ever-increasing demand for goods and services and the amount of carbon being discharged into the atmosphere annually is already around 10 billion tonnes[80]. See Chapter 3: Complexity.

All sections of the industrial infrastructure are expanding and among the fastest is the growth in electricity generation, motor vehicle manufacture, consumer goods manufacture and aviation. As identified in Chapter 3, estimates for increases in electricity alone may at least double by 2050.

This is against a background of emissions needing to be reduced —not increased.

Present policy settings will struggle to hold emissions at the current level. A continuance of the existing approach to business and commerce will not be sufficient to actually

---

[80] This includes land use changes and is over 36 billion tonnes of CO_2: To convert quantities of carbon to carbon dioxide (CO_2), multiply by a factor of 3.67.

reduce emissions, unless those people in the less developed parts of the world are denied the right to achieve the levels of affluence currently enjoyed by those in highly developed economies.

Admittedly, there is a redundancy in repeating the following message, but then again, it is an essential message. The research and implementation of technologies, that we can reasonably expect will be commonplace by the end of this century, must be brought forward to the first third of this century.

This is the message that does not yet appear to have adequately filtered through to all levels of business, industry, commerce and government.

It is crucial that programs are put into place to expedite the development of these future technologies and that these programs are appropriately funded.

Although the current threat is *totally* different, it is arguably greater than the threat posed by all the wars of the twentieth century combined. The human and financial cost of war is unspeakable, so it should not be too difficult, to imagine the cost to humanity, if we continue to war with our own environment—and the environment wins!

The human race should be careful not to trigger a crisis in the biosphere and test the veracity of an argument, occasionally mooted, that human civilization does not respond adequately to challenges to the social order.

The global conversation must begin to set priorities that will ultimately lead to restructuring the way resources are used, which may also change what we believe about existing economic paradigms. We must not hold back on our defence of this planet's environment.

Maybe the best perspective we have on the wonder and fragility of our environment is garnered from outside and the only 'outside' available to us is from space. Powerful commentary is available from those who have pondered, studied or travelled into space and may provide those of us who remain 'Earthbound' with an essential roadmap. To save our planet, we need to heed the advice of those brave souls that have seen what most of us will never witness first-hand. We should not hold back in our defence of the Earth's ecosystems.

12. Science and Engineering — The Future

It is likely that existing technology will be inadequate to provide the energy, food, materials and water required by a world population of over nine billion by 2050. Therefore, we need to bridge the technology gap and simultaneously solve the highly complex problem of a changing climate. Achieving this depends on public opinion being positive to taking up the challenge. Also, political and business leadership must be convinced of the need for action, on a scale comparable to the effort demanded by WWII. The scientific and engineering solutions will emerge as new economic and funding models are promoted.

The third critical climate protection measure identified in the introduction urges:

'Expedite development of future technologies. There will be a profusion of zero emission fuel, energy, transport and manufacturing capability by the end of this century, but this may be too late. Establish a goal to implement the non-carbon energy systems of the latter half of this century to 2030 or earlier. Give this the highest possible policy priority.'

Right now, there appears to be some reluctance in mainstream political, scientific and engineering circles to consider the possibility that our current science and engineering capability may be inadequate to meet the challenge of global warming. This is in marked contrast to the environmental scientific evidence, which clearly indicates dire ramifications of continuing with the existing business and resource management practices based on current science.

Perhaps there is wisdom in avoiding any suggestion that our current scientific and engineering capabilities may not be adequate to meet the challenge of global warming. Mainstream science and governments around the world could be choosing not to alarm the population with the slightest hint that at this time we may not have the means to preserve our environment.

Of course, it is also possible that not everyone's number one priority is the preservation of the environment. Especially if it requires a reduction in some of the excesses that we in the advanced economies of the world have enthusiastically embraced and that keep the economy ticking along on a business and solutions as usual paradigm.

Action to mitigate global warming does not receive a very positive response in the media, or from many in the political arena. The lukewarm response in these powerful sections of the community does not necessarily reflect the view of many others, who certainly do not wish to see a situation develop where the stability of human civilization is threatened. In particular, if appropriate action is delayed solely on the grounds that people need to be protected from a disturbing reality. Surely, none of us would want to jeopardise our world and comfortable lifestyles because we were too afraid to take the hard decisions.

In today's world, catchphrases such as: 'It's all right to say that with 20/20 hindsight' or, 'It's easy to be wise after the event' are frequently repeated. It would add insult to injury, if we find ourselves in an irretrievable atmospheric situation and also have to endure statements such as 'Why wasn't all this looked into before—when we still had time to change things?'

High-level research into climate change mitigation is the challenge of our age.

The Challenge

Satisfying the energy and resource needs of a burgeoning industrial sector as it responds to the upsurge in consumer demand, is one of the world's biggest challenges. These energy demands must be met without the vast CO_2 emissions associated with coal, oil and to a slightly lesser extent, natural gas. This task is beyond current technology or we would be already doing it. It is also beyond current scientific and engineering research—or we would be starting to see viable solutions emerging.

It is probably not realistic to consider that current research on carbon capture and storage (CCS) will provide a practical method of containing all the CO_2 from the world's existing coal, oil and gas fired power stations, let alone those being built or planned, to cope with the ever-increasing demand for power.

'Extreme Science' – the Solution

The concept of 'extreme science' borrows from that of 'extreme sports'. Extreme sports push every boundary of skill, endurance, daring and risk control to the edge of what would normally be considered prudent.

In the context of research into mitigation of global warming, extreme science would push every scientific and engineering boundary to the limit. However, there is *one important distinction between extreme sport and extreme science and that relates to risk.*

In the case of global warming, risk levels will be pushed to the very edge of control if we do not challenge the boundaries of science and engineering.

Thousands of examples of extreme, or high-level, science exist right now; one such example is the advanced scientific research being undertaken with the Large Hadron Collider, situated cross-border between Switzerland and France. The Large Hadron Collider is used for high-energy particle research and is just one of the high-level scientific research establishments around the world.

In the main, these establishments do not target the decarbonisation of our energy and fuel systems. Research to decarbonise the economy is not commonplace, and commonplace it must become. The twin tasks of stopping the accelerating rate of atmospheric CO_2 and also ceasing to compromise the capacity of the oceans and forests to act as CO_2 sinks, will not be easy. The challenge will be to minimise risk. However, the risk and danger of not engaging in substantive action will almost certainly incur a far higher cost in terms of our future well-being.

Mitigating the worst effects of global warming requires higher expenditure on scientific and engineering[81] research and development, than at any previous time in history. The most significant and massive scientific and engineering projects in the past have demanded considerable capital expenditure.

These projects include the Panama and Suez Canals, Brooklyn Bridge, Channel Tunnel, Hoover Dam, Three Gorges Dam, Apollo Moon Missions, Manhattan Project and many others. For each of these projects, as a percentage of the gross domestic product of the time, the expenditure was large. Funding the global warming mitigation project will probably be very much larger.

The total world expenditure on financial rescues, bailouts and various stimulus packages[82], provide an indicative benchmark of possible funding levels. The collapse of our relatively benign biosphere would be a far more serious threat for people than the collapse of a financial system. These monetary systems are a human construct that are managed by humans. The biosphere does not answer to us for its response to our inputs.

We should not be deterred by the magnitude of this challenge. We need to fine-tune our thinking and develop a mindset that reframes the momentous changes needed to ensure and insure our way of life, as challenges—not threats.

---

[81] Science provides the way and engineering provides the means.
[82] Years: 2008 to 2015 and counting.

Change is not the enemy.

Meaningful and purposeful debate on whether society is prepared to commit to funding the science and engineering required to support much needed advanced technologies must start now.

Currently we are locked into hydrocarbon fuels. With current engineering, there are no practical alternatives to do the 'heavy lifting' of manufacturing, transport and electricity generation, on a global scale. It is therefore imperative to expand our horizons from the fuels that have been used for hundreds and in the case of wood, thousands of years.

Most sources of energy such as wind, water and the sun have been used throughout human history. With the possible exception of nuclear fission, photovoltaic cells and similar sources, there are presently few completely new methods of obtaining energy.

Modern humans of course, have far superior technologies for burning hydrocarbon fuels than existed in the past. Even so, underlying all these fuels is the same original fossilised carbon fuel source. Wind, water, geothermal and the Sun have not been forgotten; however, they do not currently provide nearly enough base-load energy for modern society.

In the ultimate sense, all energy and life, derives from the Sun. There is no 'new' energy, only new or different methods of converting the Sun's energy into forms we can readily use.

Where to for Science and Engineering?

It is impossible to predict the exact outcome of research that has yet to be endorsed, funded and undertaken. However, it is reasonable to speculate that with the vast amount of solar energy available to the planet every day, the Sun is likely to underpin the world's long-term energy requirements.

In the main, the areas of the world that are subjected to the most intense solar radiation do not coincide with the areas of highest energy demand. Therefore, part of any solar solution will involve efficient transmission of electricity in 'base-load scale', from high solar intensity areas to the world's major industrial and population centres. This is where advanced science and engineering is essential.

A crucial science and engineering question is: How does the human race harvest energy from the sun in real-time, from areas of high solar radiation and distribute it to the far reaches of the globe, again in real-time, with minimal energy loss? Development and deployment of very advanced methods of capturing and distributing solar energy are essential.

The techniques for this kind of engineering could be described as 'advanced solar technology'. One example of this may be the development of high temperature current-stable superconductors, capable of transmitting base-load energy with negligible voltage drop[83].

There are other possibilities. One possibility is the development of long distance power transmission without physical conductors, using devices that are in the early stages of conceptual imagination and may not yet have even reached the 'drawing board' stage. Another is the possible scaling-up of superconducting technology to enable power to be transmitted over transcontinental distances.

If superconductors can be developed that are able to operate with vastly greater stability than available today and at the same time carry substantial and rapidly fluctuating currents at global ambient temperatures[84], then the massive solar thermal plants of the future, as yet only imagined, may be the solution.

---

[83] Negligible voltage drop is the principal characteristic of a 'super conductor'.

[84] A single global ambient temperature is not meaningful in the context of power transmission, but it is used to illustrate that in consideration of where superconductor technology is today, the use of superconductors for the purpose described in the main text will clearly fall into the area of 'extreme science'.

The 'yet-to-be-imagined' energy harvesting and distribution systems will follow the 'yet-to-be-imagined' funding of 'yet-to-be-imagined' science and engineering

A key premise underlying this book is that given adequate investment, completely new methods of harvesting and distributing energy will be developed. These may include civilian use of nuclear fusion or some of the energy sources and distribution concepts mentioned above.

Cost of 'Extreme' Science

The big question is how much to spend on research to speed up deployment of future technologies. The amount expended to rescue the world from the financial abyss in 2008/09, would make a good start. As argued earlier, it does not seem excessive to consider that such an amount could be spent on stabilising the chemistry of the atmosphere. After all, a rogue biosphere is potentially far more dangerous than a rogue financial system—despite what the banking industry may tell us.

So what has the financial crisis and its aftermath cost so far and how much more will it cost? It is doubtful that the full cost will ever be exactly quantified on a global basis for numerous reasons. A significant hurdle to an exact calculation is that in 2015, in some parts of Europe and elsewhere, the cost of adverse outcomes from the 2008 financial crisis are still being experienced.

The estimated cost appears to depend greatly on what is included and excluded from the costing. If immediate and ongoing output losses are included in the estimate, together with the stimulus packages and under-performing loans, the amount is increased by the equivalent of trillions of (US) dollars.

It is very doubtful that the world escaped with a cost of less than $US15 trillion and if ongoing losses of output and other 'downstream' effects are included, then a total cost of $US50 trillion is not out of the question.

The cost of the financial crisis is extraordinary; we must do better with the environment and avoid the expense of an unstable biosphere.

The final chapters of the financial crisis are still to be written and the total cost yet to be determined. Regardless, it is certain that an enormous amount has been spent on mitigating this particular crisis and somehow the money has been found to do this. One cannot help but suspect that if the finance industry collapses again, it would once again be bailed out. The finance industry appears to enjoy the ultimate 'sacred cow' status, whereas that privilege should really be reserved for the biosphere.

Right now, the priority ought to be the stabilisation of the chemistry of the biosphere. Currently there is little indication of a workable, scalable system to make this happen.

It is not implausible to imagine that with appropriate funding, coal, oil and natural gas may be superseded within a decade or two. This is not a 'pie in the sky' concept. A mere twenty years ago, the functionality built into today's smart phones would have been just a dream. Also, if we delve into history, within a few years of space travel being in the realm of science fiction, Neil Armstrong and Edwin Aldrin voyaged to and walked on the Moon.

Space travel, with the technology available in the 1950's and 1960's, would have seemed almost impossible at the time. We now have much greater technology at our disposal and the elimination of hydrocarbon fuels from the economy is another challenge we can conquer.

In the future, management of resources must include management of the atmosphere as a CO_2 sink because the ability of the atmosphere to absorb CO_2 is rapidly emerging as the world's most critically scarce resource. The funding implications and related matters such as possible new

130

economic and resource management models are explored further in Chapter 14: Funding the Future.

Managing and Achieving the Change

If we collectively make the decision to drastically reduce anthropogenic greenhouse gases, it will initiate change on a scale more far-reaching than most of us have contemplated. The magnitude of the change does not have to cause a reduction in living standards, or lead to a less pleasant lifestyle; this will depend on how the change is managed and implemented.

Management of the change will challenge everyone, particularly those with political and corporate leadership responsibility. Some professions that are currently not in the forefront of economic activity will become more important and valued, with remuneration levels in those professions increased accordingly. The best minds must be attracted to careers in science and engineering; both being areas of expertise that will require exponential expansion.

Leaders in banking and commerce are rewarded with salary packages that have equivalence across the globe. The same rules should apply for those working in fields such as physics, chemistry, mathematics, astrophysics, nanotechnology and also mechanical, electrical, electronics, civil and climate engineering.

New Horizons for Scientists and Engineers

Individuals working on climate stabilisation may need to attract a higher remuneration than anyone else. The very best will need to be recruited; they will be working on the most important projects in history. Financial rewards should be set such that the top graduates, from the top colleges and universities, look more favourably on degrees in the sciences and engineering, than in business and commerce.

The team gathered for the Manhattan Project at Los Alamos in New Mexico in 1943 may provide a template for how priorities can be set, if the task is considered urgent enough. Infrastructure development will be a key part of firstly holding and then reducing, greenhouse gas levels in the atmosphere. Scientists and engineers may not only be the new captains of industry, they may also have a significant role to play in the politics of climate change.

Surely no task can be more urgent than preserving our civilization.

New Problem – New Thinking

Today's problem is new, and unprecedented in the entire history of human civilisation. We can no longer base our economy on burning the carbon-based fuels that have been the mainstay of our anthropogenic energy sources since the Stone Age. This new problem necessitates a new worldview and the one thing of which we can be sure, is that our future safety, security and prosperity cannot be based on the old carbon fuel-burning worldview.

Burning carbon-based fuels, whether fossilised or otherwise, is what created the problem of excessive atmospheric CO_2 in the first place and continuing with more of the same, will never solve the problem. Any future increases in the efficiency of burning fossil fuels, will be swamped by the increased demand as people in less developed economies catch up with the energy demands of the more developed economies.

Expanding on this idea, while CO_2 emission limits, carbon taxes and trading schemes are essential in transitioning away from fossil fuels in the short term, the long-term solution must be found in the deployment of completely new technologies to satisfy our energy needs. Earlier in this chapter, it was strongly suggested that any new energy technologies would utilise the plentiful solar energy the Earth receives every day. The Sun has been fuelling the bulk of our energy requirements, including keeping us warm and growing our food, since time immemorial.

The solar solution will have three components: storage, portability and transmission of solar generated electricity over vast distances with negligible voltage drop. This will not be impossible.

The world cannot give up on science and engineering at this critical point in our history. Because fossil fuels have been the principal source of energy[85] to power the Industrial Era, it does not follow that we should still be burning coal and oil, as the principal source of energy in 2015, let alone in 2020 or 2030. We must be confident that new and as yet, largely undiscovered technologies will provide the means to replace carbon based fuels sooner rather than later.

---

[85] The energy contained in fossil fuels originally derived from the Sun. This solar energy was geo-sequestrated millions of years ago.

13. Universal Carbon Cost

Until now, humans have really had fun with the environment[86]. We have built stuff, made stuff, zoomed around all over the place and eaten until we were replete. In addition enormous amounts of energy and resources have been devoted to the many wars that have blighted human history. All this has been done with scant regard for the environment, because nothing seemed to run out. If a commodity became scarce, we just went out and found some more and if that didn't work, we found a substitute.

It seemed that the party would never end. That was until we started to realise that the massive quantities of a colourless, odourless gas being released into the atmosphere by industry, business, transport and agriculture, was not entirely harmless to the environment.

This gas is carbon dioxide—CO_2.

CO_2 is not one of those commodities that is about to become scarce. What is becoming scarce is a large enough carbon sink. Even the Earth's atmosphere is not large enough to absorb all of our CO_2 without ill effect. Of course, it is not just CO_2. Other greenhouse gases are also destabilising the chemistry of the atmosphere. However, it is the massive quantity of CO_2 in particular, that is most rapidly tipping the system over the edge.

The party is over.

Cradle to Grave

'Cradle to grave' or 'whole of life' product cost is the total cost of each thing we consume, or is consumed on our behalf.

---

[86] Unfortunately this does not apply to everyone on the planet, but it certainly applies to those of us who have done most damage to the environment.

'Whole of life costing', including a realistic cost of CO_2 emissions that result from all stages of the manufacturing chain, is essential public policy.

This costing should include: Manufacture of the product, energy and other operating costs during the life of the product, recycling and the ultimate disposal of the product. Such costing decisions must become the basis for shaping future policy on the manufacture and distribution of all goods and services in a carbon-constrained economy. Implementation of a proper and fair 'whole of life costing' system ought to be considered by society to be a challenge and opportunity, not a burden.

Carbon Cost of Everything

Everything really does mean everything, even the relative greenhouse gas cost of our final journey. For anyone who is concerned about our lasting footprint on the planet Earth, how do we currently compare whether cremation or burial leaves the smallest CO_2 emission signature? Where is the line between cremation, burial in single person graves, two person graves and vertical graves?

This example may be challenging because the end of life is a difficult time. There is no suggestion that the State or any other party should control a person's final resting place. However, death happens to us all and people should be able to make an informed choice. I have listened to a number of people, who have accepted cremation as the norm for many years, now ask: 'But what about the greenhouse gases; maybe I should be buried instead of cremated?'

The real point is that most of us haven't got a clue about the ultimate greenhouse gas emissions associated with anything we do or buy.

With more than seven billion people calling this planet home, we cannot afford to waste a single kilogram of unnecessary CO_2 into the atmosphere. Of course, imagining that we shall ever be able to actually account for every last kilogram of CO_2 emitted into the atmosphere is unrealistic. But right now, with only limited specific and accurate data, economic policymaking is significantly compromised.

Assessment of the environmental impacts associated with all stages in the lifespan of major infrastructure projects, buildings and manufactured products is not a new area of study. The first analytical systems were developed in the 1960's and 1970's. Among the environmental impacts considered today are land, water and atmospheric pollutants and, of course, carbon dioxide.

Any pollution has some potential to harm the environment, so why is there a need for a special focus on a universal carbon cost and for a carbon cost of everything? Carbon dioxide is different; CO_2 emissions into the atmosphere are a separate issue.

The CO_2 that is being released into the atmosphere today is either retained in the atmosphere or absorbed by the oceans and was originally part of a sequestration process dating back millions of years. Anthropogenic carbon dioxide emissions must be separately costed and the cost to the environment taken into consideration. The topic is further discussed in Chapter 14: Funding the Future.

There is convincing evidence of the need to consider CO_2 as a special case. In the concentration levels found in the atmosphere today, CO_2 is still harmless to humans and animals. However, there is a very important proviso, which concerns the billions of tons of CO_2 that continue to be discarded into the atmosphere each year as a by-product of combustion.

It is difficult to obtain exact numbers, but a conservative estimate is that the amount of CO_2 dumped into the atmosphere annually far exceeds the total quantity of industrial and domestic waste tipped into landfill or recycled, in the same timeframe. Additionally, it is important to realise that the process of recycling often involves even more CO_2 emissions—especially the process of 'hot/remelt' recycling. No process that involves fossil fuel generated heat is a one hundred per cent zero sum game.

There are problems associated with any waste disposal. However, we do have the choice to either recycle, or establish new landfill sites for solid waste; we cannot recycle or find a new atmosphere.

The urgency to reduce emissions means that collecting data about CO_2 emitted in the provision of every good and service used by the world community cannot be deferred for much longer. This information will not, of itself, compromise overall employment. What it will do is change the nature of some work as we take into account additional and different inputs into the economic discussion. The world has the capacity to do this job; it has the people to do it and it has the computing power.

An examination of the great sweep of human history indicates that information promotes free societies, just as lack of information empowers oppression.

Therefore, before images of a big brother society spring to mind, we should remember that information is simply information and in this case it is crucial, if anyone at an individual, corporate, national or international level is to make sensible decisions concerning the stability of the biosphere. The cost of emitting carbon dioxide into the atmosphere must be calculated. This cost must be factored into national budgets by society as a whole and the funds generated used to underpin research into the new economic paradigm that will secure our collective futures.

Limits to our Current Models

Current models for managing resources, energy, agriculture, water and industrial capacity are based on and require exponential growth in goods, services and money supply. While this may not be the official economic model, in reality it is. Witness the tracking of the economic growth figures for all countries; continuous growth is good, contraction of growth is bad. Of course, there will be growth in the world economy as living standards are raised, which particularly applies to those people and countries, currently coming from low economic bases.

An economy that is not dependant on perpetual growth is a challenging concept, but it will be a key to sustainable living in a carbon constrained world. There will be a plethora of entrepreneurial opportunities as the economy is re-engineered from a carbon to a non-carbon energy base. There will still be money to be made!

The default status quo of exponential growth has been with human society from time immemorial and has worked exceptionally well in raising living standards and comfort levels for hundreds of millions of people. Nevertheless, in a world of finite resources, including the atmosphere, exponential growth must have a limit somewhere or sometime and it is just possible the human race is close to finding it, unless we find a way of resetting the energy equation.

Current economic models do not accurately reflect the true cost of many economic activities as they do not take into account many externalities[87]. A significant externality, not costed in the perpetual growth model, is the use of the atmosphere as a CO_2 emissions sink. This un-costed externality was unknown and took us by surprise until sometime between the 1960's and 1980's.

Damage that the Industrial Era has caused to the biosphere in general and the atmosphere in particular, needs repair and there will be no *silver bullet* to carry out this repair that is so vital to secure our collective future. Human action alone will not be sufficient to fix the biosphere. What humans

---

[87] In economic terms, an externality is a cost or benefit of a transaction that is not reflected in the price of the transaction and is carried by a party or parties not directly involved in the transaction.

must do now is move as quickly as possible to avoid the continued break down of chemical systems that have been relatively stable for hundreds of thousands of years. If we can do this, then processes within the biosphere may slowly restore the balance. There are no absolute guarantees, but if we don't start soon, it will only get worse.

Tracking CO_2 Emissions

Solutions will not be found without including the cost of the effect on the biosphere of human industrial activity; a cost that is currently external to the normal commercial and industrial costing systems. The environmental impacts (externalities) associated with using the atmosphere as a zero cost CO_2 dump cannot be fully calculated without first knowing the CO_2 emission cost of everything we use, in kilograms of CO_2 emitted.

We are familiar with the concept of 'food miles'[88] and the fact that more emissions are associated with a food item that has travelled half way around the world than with locally grown food. To enable this concept to be a more precise decision-making tool for food selection, *the emissions associated with the production of the food should also be known.*

Food production emissions will not be identical between different production locations around the world. Other factors will need to be taken into account such as the mode of transport (e.g. air, sea, rail, truck, canal, or even on the back of an ox) and the cost, in kilograms of CO_2, associated with each of the transport modes.

Knowing the CO_2 emission cost of everything will mean that as well as food miles, we shall have 'electric toaster miles', 'table and chair miles' and 'new shoes miles'. This will permit a true comparison of emissions between all goods, as well as

---

[88] The distance an item of food is transported between the point of origin and the consumer.

allowing comparison of those goods that have travelled halfway around the world and those produced locally.

There will be speculation about the effect this may have on the balance between local and distant manufacture and this effect will be determined over time by consumer and strategic choices. Currently the shipping and other overhead costs that are added to the actual cost of manufacture of a product are effectively based only on the monetary cost of providing the particular overhead.

Therefore the cost of manufacture and the overhead component of any equipment, product, machinery or service will vary markedly from country to country. This can enable goods originating in a distant land to compete on a basis that may not have been the case if all CO_2 emissions had been priced and considered.

However, if there is an agreed cost to emit each tonne of CO_2 into the atmosphere, across all countries, one competitive advantage will depend on reducing the CO_2 emissions per unit of GDP[89]. This will hasten the race to 'decarbonise' the world economy, which should be the overwhelming priority for us all.

When total CO_2 emission costs are calculated, *including a proportion of the emission cost* of building ships, providing crews, fuel, docking facilities, administration, marketing, office buildings and other infrastructure, the overall comparison equation between 'distant' and 'local' may change markedly. The price of using the atmosphere as a carbon dioxide sink must always be factored into the total cost of any good or service.

For a realistic 'local' to 'distant' manufacturing comparison, the inputs must be expanded to include the greenhouse gas efficiency associated with the each manufacturing plant or manufacturing process. The relative greenhouse efficiency of inputs to the manufacturing plants

---

[89] This is frequently called the 'emissions intensity'.

such as electricity, gas and water must also enter the calculation to determine the final cost of the CO_2 emissions.

Only then will we get a clear idea of whether it is cheaper to manufacture a product or harvest and process a crop, in a distant or local manufacturing or food processing plant. The world already has the fast computers, specialist software developers and other people who are more than capable of managing these tasks.

Expanding on and adding detail to the concept of universal carbon costs, imagine an electric drill, one kilogram of iron ore, a steak and a smart phone. There will be a total CO_2 and other greenhouse gases emission cost of delivering those objects into the consumer's hand.

For the electric drill, it will include the CO_2 cost of the iron ore, the cost of refining and producing the steel components from the iron ore, the crude oil and the cost of refining the oil and producing suitable polymers for the plastic components. It will include all of the other materials and components that are required for manufacture and also the total cost of manufacturing, shipping, marketing, distribution and retailing.

Considering the one kilogram of iron ore, it will be the emission cost of mining, explosives, dump trucks, hard hats, aircraft and fuel to fly-in/fly-out the labour force and shipping to the ultimate manufacturing plant. There will also be the CO_2 emission cost of marketing and administration, including a share of the necessary infrastructure and built environment.

The CO_2 cost per kilogram of iron ore would be small compared to all the refined, manufactured and assembled steel, non-ferrous and polymer components within an electric drill—taken on a per kilogram basis. Similar, but different processes would apply to the steak and the smart phone and in each case, the process will be complex.

There will obviously be variance between methods of *producing the same product in different factories in different locations.* Taking an example from just one industrial sector, the production of aluminium, CO_2 emissions per kilogram will vary due to many factors and the variance may be large. Key variables will include the actual processes used during the refining, smelting, ingot plate and sheet rolling and rod and wire drawing.

These processes will involve heavy ore processing machinery, furnaces/smelters, rolling and drawing machinery. The bauxite must be mined, alumina refined and shipped before any of the above occurs. There is also a range of energy sources that may be used to generate the electricity used in the above processes. This difference will allow lower CO_2 emission options to be fostered.

The CO_2 cost would probably be charged at the point of sale, similar to a value added or goods and services tax. If part of the CO_2 charge has been paid at an intermediate point of sale, for example on bulk iron ore, coal or crude oil, then CO_2 credits will be deducted from the point of sale total that would normally accrue later in the supply chain. The audit trail for CO_2 credits will be exacting, but not beyond the capacity of the computer networks and intelligent software available.

There are many examples of existing complex tracking systems. One such system is used to track aircraft components for compliance to quality standards. This includes tracking the results of the mechanical and other testing procedures, of all material used in the manufacture of component parts. The particular batch of material used to manufacture specific component parts can be tracked and audited. Validation of testing at all stages of production, hours of service and modification records can all be tracked.

The compliance/audit trial for CO_2 charges will scarcely be more complex, even allowing that the emission charge may be paid at any stage of manufacture or processing. The cost of CO_2 emissions could be applied at any time in the manufacturing cycle, such as when a bulk commodity is shipped, a consumer product manufactured, or processed food passes from one supplier to another.

Most, if not all, of the numbers that are necessary to set prices for CO_2 emissions, for every facet of our 21st century lives are probably available, but they are currently scattered over such a large number of disparate databases as to make collation almost impossible—until it is made a worldwide priority. Data is collected by industrial, commercial, transport and shipping groups, also by government, industrial and privately funded research establishments, universities, the military, and intelligence agencies.

A coordinated effort to establish a carbon emissions accounting system could draw from and build upon, this existing information; most of which is currently not accessible to the world at large. Even the reach and capacity of existing Big Data/Data Analytics models may need enhancement to manage the massive volume of data that will be required to implement, interpret and disseminate information from such a large and disparate system.[90]

Implementing a Universal Carbon Cost

The implementation of a universal carbon cost and the actual form of an emission pricing system can evolve together. The ultimate pricing system will be a policy/political decision and could take the form of, for example, an emission cap and trading scheme, or a carbon tax with or without border adjustment taxes and tariffs.

While the set price should be no higher than is necessary, the price applied will initially need to be upwardly progressive; sufficient to reclaim the chemical stability of the atmosphere within a timeframe sufficiently short to make a difference; within one or two decades rather than three or four. Complicating factors such as how land use[91] is worked into the pricing structure will also need to be resolved.

---

[90] 'Big data' is a term used to describe data-sets that are so large and/or complex, that usual methods of data processing are insufficient.

[91] Agriculture, forestry and land use for production of bio-fuels.

If a country decides in its own domestic market not to participate in a carbon-pricing scheme, it will be problematic for global emissions reduction. How problematic will depend on the size of the industrial sector within that particular country. However, as most countries rely heavily on international trade, this problem could be mitigated by an importing country applying at their border, an estimated carbon emission cost of manufacturing in and shipping from, the originating country.

CO_2 emissions 'leakage' to non-participating countries, can be addressed if the final CO_2 price is levied on all products, either at point of sale or at appropriate points in the supply and transport chain. Depending on the details of the final emission price, border adjustment taxes or tariffs may be necessary. A carbon price can then be applied, whether the product originates next-door or 12,000 kilometres away. This price will include the CO_2 emission cost of everything: resource extraction, refining, shipping, manufacturing, transport of finished goods, marketing and retailing—a system that will be equitable for all product originators, whether local or distant.

Importantly, the CO_2 emission charges that are levied at point of sale or border will be available to the government of the country levying those charges, with allowances made for any amounts imposed during the manufacturing process. These CO_2 emission charges should not be considered an *additional* levy, or barrier to trade, if they form part of a legitimate taxation system and trading tool.

Political agreement between governments on the cost of using the atmosphere as a carbon dioxide sink must precede any new agreements that would necessarily be made under the auspices of the relevant international bodies and trade agreements.

Currently, there are many differing estimates for the cost per tonne of emitting CO_2 into the atmosphere. The only certainty is that the longer we delay, the higher will be the cost, irrespective of whether the final pricing depends on 'cap and trade' emission trading schemes, with or without border tax adjustments—or some completely different economic mechanism.

Taxation is discussed further in the next chapter.

We should all look forward to the day when it becomes accepted that a rigorous worldwide carbon pricing scheme, fair to both rich and poor nations, is recognised as necessary and is implemented. The task ahead is obviously more complex than outlined above, but here we are in 2015, still with no worldwide agreement and scarcely the faintest chance of arresting the level of CO_2 in the atmosphere before it reaches 450 ppm, which is likely to create a *carbon dioxide equivalent* (CO_2e) close to 500 ppm. It is well past the time to begin this important work.

14. Funding the Future

Bold economic leadership is essential to mitigate global climate change before climate misfortune and economic disorder begin to overwhelm the planet and its governments. As we know, there is always the ability for governments and corporations to prioritise expenditure for projects deemed as urgent.

We are now at a time in history where putting money away for a 'rainy day' becomes meaningless as the rainy day is here. Before the 2008/9 global financial upheaval, Australia—a mid-size economy—had approximately $A90 billion in future funds and projected budget surplus.

As an example, if Australia in 2008 had allocated $A20 billion, $A30 billion, $A40 billion or even $50 billion and had explicitly earmarked it for research and development (R&D) into *zero emission energy sources and fuels*, the complete transformation of the economy of the world from fossil fuels to non-polluting sources, could be measurably closer.

International competition and free enterprise would ensure that funds committed, in good faith, to this cause would soon be matched and surpassed, by many others. Also, as a bonus, many years would elapse before the full financial commitment would be required. An example of this is illustrated by the 'Snowy' scheme, as detailed later in this chapter.

Industrial research organisations that are directly funded by governments will have a key role. However, in accordance with current economic paradigms, existing universities and the private corporate sector would necessarily underpin much of the expansion of R&D facilities.

Seven years later and the potential for this is history and in 2016, if nothing changes, the same will be said about 2015. It is possible that the many billions spent on financial stimulus since 2008 could have been more effective, in both the short and the long-term, if targeted to the building of research capacity to decarbonise the world economy.

It may also have been more cost effective than the programs that were actually implemented. Imagine if even half of the 2008 to 2012 global stimulus spending had been quarantined for projects to accelerate research on zero emission technology. The difference to the task of expediting the science and engineering of the future could have been immense.

'Crying over spilt milk' and lost opportunities, is futile and a waste of time, unless of course it stimulates a reassessment of priorities in the future. No single country, including those with the largest Gross Domestic Product (GDP)[92] and/or populations, will halt the warming planet if acting alone.

However, if one or two governments take a strong lead in expediting future science and engineering, with the aim of eliminating fossil fuels from the energy, manufacturing, commercial and distribution sectors of the economy, others will follow.

It is not too late to do this. With adequate resources,
dreams do become reality.

Implementation Models – Past & Present

The best example of how effectively the human and material resources of a country can be rapidly mobilised and reallocated, is manifested in the worst of circumstances—war. In times of military conflict, all countries will mobilise the resources necessary to conduct campaigns of epic proportions.

---

[92] Or, Gross National Product (GNP); frequently there is only a small difference in the final numbers.

Just imagine the total military spending during the Second World War. As a small boy during this time, living in one of the combatant countries, it is my recollection that there was still money for food, medicine and other essentials. More importantly, after the expense of six years of total global war, money was still found to rebuild the economies of both Allied and Axis countries within—relatively—only a few years.

Please do not tell the citizens of this great planet that we cannot afford to eliminate fossil fuels from the world economy starting with a 5% reduction in 2016 and a further 5% each year after that. It is all just a matter of priority.

In terms of today's money and Gross World Product, the expenditure by all combative countries in WWII would (today) be equal to many trillions of dollars. The world must unite and *declare war on the use of hydrocarbon fuels to underpin the economy.*[93]

Elimination of fossil fuels may end up being as expensive as WWII. This should not be viewed as outrageous expenditure. Runaway climate change has the potential to be an even greater threat to humanity than the very *dangerous state of affairs that existed during all the wars of the 20th century.*

Currently, there is not any climate specific funding, or commitment of resources equivalent to the financial and resource commitment that was required during those periods of conflict.

During major world conflicts in the past, previous generations were not found to be wanting. The least our generation can do is adopt a positive 'can do' attitude, to the financial measures that will be required to reverse the rate at which atmospheric greenhouse gases are increasing.

---

[93] The complex relationship between aviation, GHG emissions and societal expectations are discussed in Chapter 16 - Sustainable Manufacture.

The 'Snowy'

In 1949, Australia commenced the Snowy Mountains Hydro-Electricity and Irrigation Scheme. The management of this project offers a financial model of how large percentages of GDP can be committed to a project without incurring ruinous financial consequences. The project is universally and affectionately known in Australia as the 'Snowy' scheme.

In July 1949, contracts were drawn up for work worth £A200 million ($A0.4 billion) to develop the Snowy Hydro-electric and Irrigation Scheme when the GDP of Australia was approximately $A5 billion. At the time $A0.4 billion represented about 8% of GDP. The entire project was expected to take 25 years and the estimated total cost to be approximately twice the original $A0.4 billion investment.

Several aspects of the Snowy Mountains scheme are important:

- it was visionary and took courage

- in 1949, Australia did not have the labour force or the expertise in deep hard rock tunnelling. Both these issues were addressed and solved with the workforce[94] being drawn from thirty different countries

- the complete project took 25 years to complete at the cost of roughly $A860 million ($A0.86 billion), on time and within 7% of the original 1949 estimate.

The size of the commitment, as a percentage of GDP, did not break the country; in fact it could be argued that it was the beginning of a 'golden age' in Australia's history. In 2015, 8% of Australia's GDP of just over $A1,500 billion would be $A120 billion and the equivalent of the total Snowy commitment would be approximately $A240 billion.

---

[94] From 1949 to 1974: 100,000 people worked on the 'Snowy'.

Imagine what could be achieved if just one country had the political determination to sign firm contracts for $A120 billion, for global warming mitigation research projects and action, with an ultimate cost of $A240 billion. This is 'food for thought', especially as there are over a dozen countries with a GDP greater than the mid-sized economy in this example.

Transposed to the world stage, where the Gross World Product (GWP) is in the order of $US100 trillion, 8% of this amount would be a little over $US8 trillion. In 1949, the 'Snowy' project with an initial commitment equal to 8% of GDP was financed by the Australian government without harming the economy.

Of course, there are hundreds of similarly massive projects around the world that were huge in relation to the GDP of the particular country, which have been successively undertaken and completed without destroying the economy of that country.

Other models for immense global projects, or examples of international cooperation that have been successfully completed, include the Apollo project to land a man on the Moon, the International Space Station, the Aswan High Dam, the Three Gorges Dam and the Manhattan project to develop an atomic bomb[95].

Importantly, these projects all had one thing in common and that is the backing of a government.

Cooperative Models

When the size of a project, whether it is a war or a man on the Moon, becomes sufficiently large, then it extends beyond the financial reach of private enterprise. The financial crisis

---

[95] Although undertaken during WW11, the Manhattan project is separately included because of the stand-alone nature of the project, within the overall war effort.

that started in 2008 is another illustration of the principle that where there is political determination and government involvement, there can also be a massive reallocation of financial and material resources.

Getting the chemistry of the atmosphere back into balance is exactly one of these 'sufficiently large' projects.

Existing research and development facilities will need to be greatly expanded. Significant expansion of the research sections of universities, industrial and government research centres will embrace major new capabilities that are focussed specifically on developing zero emission energy and fuel technologies.

As discussed earlier, 'industrial research organisations that are directly funded by governments will have a key role. However, in accordance with current economic paradigms, existing universities and the private corporate sector would necessarily underpin much of the expansion of R&D facilities.'

All of these establishments and organisations could be part of an *'8% of GWP program'* to stabilise the chemistry of the atmosphere. If a 5% emission reduction is planned for each year, starting in 2016, the job should be largely completed by 2035.

Over the past couple of decades there has been no CO_2 emission mitigation expenditure anywhere in the world sufficient to significantly cut rates of global emissions, let alone reduce the overall atmospheric total.

This is evidenced by the, still rising, 400 ppm of CO_2 in the atmosphere. Development of new technology is required across the board; for power generation and transmission, global transportation, manufacturing methods, the built environment, agriculture and land use.

Seed funding of 8% of Gross World Product (GWP), the equivalent of approximately eight trillion US dollars, should be committed, right now, on a worldwide basis. This will substantially fund the scientific and engineering research necessary to develop the technology required to change from business and solutions as usual to sustainable future models.

This will include bringing forward projects that are designed with the express purpose of developing new and exotic fuel and energy sources; energy sources that otherwise, would probably not be commonplace until the end of the century. A further 8% of GWP ought to be available as contingency commitment.

In recent times there has been much discussion of setting aside monies for the explicit purpose of helping developing countries to adapt and alleviate the harmful effects of climate change. This may well be necessary, but preferably there will be a concerted and focussed effort to eliminate, as far as is possible, the need for such adaptation.

If we wait until the physics of the Earth's climate catches up with the scale of mankind's changes to the chemistry of the biosphere, it is doubtful that even 8% of GWP will be sufficient for greenhouse gas abatement. The 2008/9 financial crisis illustrated how rapidly costs can escalate. That particular crisis was triggered by the financial markets in just one country. The costs associated with a full-blown, planet-wide environmental crisis would be impossible to estimate.

If humans can muster the required vision and drive, imagine the progress that could be made to develop enhanced base load solar power generation systems, minimal voltage loss electricity transmission systems, fuel cells, solar desalination plants and other low emission technologies not yet imagined.

Our insatiable human appetite for more and more material goods must not be forgotten and we may not always be dependent on iron, steel and hydrocarbon based polymers for these material goods. However, the Earth's biosphere cannot wait until the end of the century for new and exquisite methods of manufacture.

Additive manufacturing (3D printing) is already digitising manufacture, but we shall need even more 'extreme scientific and engineering' solutions to minimise the demand on the planet's physical resources—in particular the atmosphere.

As the shared will is gathered, other funding channels will become obvious and will probably be preferable to funding by governments. As the financial industry joins with government, scientists and engineers, there will undoubtedly be any number of innovative funding schemes available.

The total amount of funding will obviously not be required for many years. In fact, it would be impossible to use more than a small fraction at the start of the project. The signaling of intent will be the most important aspect in the early stages of the commitment.

Everyone should be involved in funding the scientific research and downstream engineering[96]: corporations, citizens and governments; people in all walks of life and in all countries. How the world community responds to the changing climate in the next five years is vital. The importance of this to civilization cannot be underestimated.

Now is the time for optimism; it is the time for us all to seize the challenge and the future, with commitment and enthusiasm.

Adjusting the Economic Model

To ensure long-term climate stability, the methodology by which governments raise revenue may need to change. The reality facing all humanity is that we must adjust to living in a carbon-constrained world. In terms of this new reality, the question is whether or not the current taxation system, which is based on revenue collection methods dating back hundreds of years, still suits the needs of today's world.

---

[96] For example, close to universal involvement may be possible by means of carbon pricing systems, modified taxation systems and targeted government bonds.

As already discussed, expenditure on immediate and long-term greenhouse gas emission reduction measures, by both government and the private sector, must be massively increased to achieve anywhere close to the essential 'zero CO_2 emission society' of the future. With this in mind, encouraging such expenditure should be an obvious priority for any taxation system.

A realigned tax system that reduces reliance on wages, company profits and wealth, to focus on the cost to the environment would better reflect the new climate reality. It also follows that allowable deductions to tax payable would be similarly applied. With the tax system aligned to the needs of the environment, market forces will begin to work to benefit both the environment and the taxpayer.

Matters of taxation are exceedingly complex. The prodigious volume of legislation and regulations governing personal, corporate and indirect taxation regimes of almost any country in the world, elegantly illustrates this complexity. Any realignment of the tax system intended to better protect the biosphere, will be just as complex, but can and must be introduced.

The next section outlines an alternative tax system that has the reduction in CO_2 emissions as a first priority. This does not negate existing economic paradigms, but it does accommodate the extra information we now have regarding the environment. Such a change in the mindset of society would largely ensure real protection of the biosphere.

Carbon, Not Income & Consumption[97]

Governments need revenue and a means of generating this revenue is taxation. Historically, possession of a 'scarce resource' was a valid basis for levying a tax on that resource. In the past, these resources were wheat, livestock, currency and wealth.

Taxation was based on the amount of these resources an individual possessed. This principle has not changed, except now the scarce resource is the ability of the Earth's atmosphere to absorb CO_2 and other greenhouse gases. Accordingly, it is now appropriate that tax is levied on how much of this new scarcity is used by an individual, corporation or government.

Using the atmosphere as a CO_2 sink, can be considered as 'possession' of a part of that atmospheric capacity. The result is not unlike the need to pay a fee to dump household rubbish into landfill, as discussed in Chapter 2.

If government revenue continues to be generated in the traditional manner, then any attempt to reduce CO_2 emissions by introducing a price on carbon, through carbon taxes or emission trading schemes, is going to be very difficult. There will always be the argument made that any levy on CO_2 emissions, irrespective of the mechanism, will represent a new or bigger tax and this will meet great resistance, particularly in highly developed societies.

So, what if the starting point for discussion was a program to phase-out all existing taxes over the next 20 years—would this make a difference? Of course, no society can operate without revenue, but a graduated phasing out of existing taxes would allow room for a carbon tax regime to be introduced.

---

[97] Indirectly this will be a tax on consumption, but not tied directly to consumption of goods and services. It would be a tax on consuming the 'CO_2 and other Greenhouse Gases' sink capacity of the atmosphere. If goods or services have zero, upstream or downstream, greenhouse gas implications, then they would not incur a tax liability.

If our current models of taxation are open to change, it may then be easier to move away from the existing 'business as usual' operating models for industry and commerce.

There would need to be phasing-out and phasing-in periods to enable the tax base changes to be achieved. Within certain broad parameters, any change to the overall revenue base has to be a zero sum exercise.

As a consequence, as carbon taxes are phased in, taxes on wealth, income, company profits and also value-added taxes, would be phased out. This process would be necessary to maintain a stable revenue base. Much fine-tuning will be needed to impose a suitable penalty on excessive CO_2 emissions, as illustrated on the following graph.

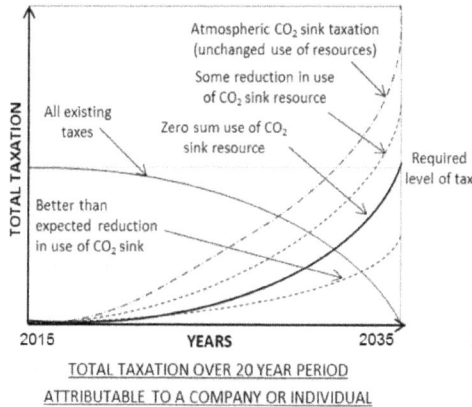

TOTAL TAXATION OVER 20 YEAR PERIOD
ATTRIBUTABLE TO A COMPANY OR INDIVIDUAL

This graph illustrates the basic principle of replacing personal and company income tax, consumption and wealth taxes with taxes based on greenhouse gas emissions. The basis for the rate of tax becomes the use of the atmosphere as a CO_2, CH_4 and N_2O[98] sink, whether for transport, energy, fuel, retail, industrial, commercial, private or agricultural purposes and processes.

---

[98] Carbon dioxide (CO_2): methane (CH_4): nitrous oxide (N_2O), respectively. For simplicity, text and graphs will frequently refer to CO_2, but all greenhouse gases would be included in any final taxation regime.

The graph could represent an individual taxpayer, company or country. It is based on many assumptions. A key assumption is that it is unlikely that any company or individual will be in a state of zero CO_2 emissions within the next 20 years.

If that unlikely event were to occur across the entire economy of an individual nation state, or the world as a whole, there would obviously be no tax base. This would be the best possible outcome for stabilising the chemistry of the atmosphere, but at the same time would obviously be an argument against complete reliance on carbon taxes. Looking at the situation from our current situation with global emissions rapidly increasing, it is hard to imagine this being a problem.

The reason for change in the taxation paradigm is that individual companies would pay less tax as the total emissions of the company are reduced. With public companies being answerable to the shareholders, paying more tax than competitors, or the industry average, should put a downward pressure on CO_2 emissions.

For individual taxpayers, their tax contribution would move from the current mix of direct and indirect taxes, to a tax regime based on the CO_2 emissions of the individual taxpayer's purchases of goods and services. However, unlike the current consumption taxes, the new tax would not be based on the value of the goods, but on the emission content of the goods and/or service. This would encourage consumers to include carbon emissions in their purchasing decisions.

Effectively nothing would change. Companies and individuals would continue to pay taxes. Payments to recipients of social security would still be assessed according to special criteria. Of course, a hermit living in the mountains would pay no carbon-based taxes, just as a person choosing this lifestyle now is paying no income or value added taxes.

The concept of a universal carbon cost, as discussed in Chapter 13, would be crucial to implement a different tax base. A CO_2 emission cost of every good, service and commodity will be needed to underpin pricing of CO_2 emissions. If the CO_2 emission cost of every single good, service and commodity in the world is not known, then CO_2 cannot be accurately or fairly priced.

As the basis of taxation moves from wealth, salaries and company profits to a CO_2 emission cost to the atmosphere, there would be activities that become 'more expensive' and others that become 'less expensive'. Overall, the switch to a CO_2 tax base should be revenue neutral or gradually reduce, as greenhouse gas emissions reduce. The purpose of this book is not to define the exact mechanism that may be necessary to achieve this, or any other system of taxation that may become the norm over the next twenty years or so.

Significant lateral thinking is warranted to address this issue.

In Summing up

There are dangers facing us every day. There is the possibility of being involved in a car crash, exposed to a penicillin resistant bacterium, or trapped in a house fire with no means of escape.

However, the universality of human-induced planet-wide climate disruption poses a threat to mankind potentially greater than any other threat, at any previous time in our history. The current global financial response is not commensurate with the risk.

We have entered uncharted and somewhat perilous waters. There is now a choice, between continuing to cling to past practices and face a potentially very difficult future, or the alternative of embracing the challenge of building a new low CO_2 emission future.

None of us wish to see the world's climate spiral out of control and deny other generations the life and pleasures enjoyed by past and present generations. We must move quickly to a new energy paradigm.

15. Gradualism, Compounding and Price Signalling

The power of compound interest, as an economic methodology to accumulate wealth, is a concept with which many are familiar. The premise underpinning this method of wealth accumulation is that by investing a relatively moderate amount of money every week and continuing the practice over a long time span, the miracle of compounding both the investment and the interest takes over and the savings are greatly multiplied.

By this method, $25 invested every week at an interest rate of 5% per annum compounding, will grow to just over $183,000 in forty years[99]. Both the regular investment and the interest components are compounding, so although the actual amount invested is only $52,000, the power of compounding makes up the rest of the money[100].

Of course, the power of compounding does not apply solely to money matters. The principle works in many areas of human endeavour as well as in the natural world and is known as exponential growth. This growth occurs whenever the growth rate of a mathematical function, commodity, organism, or population is proportional to its current value, amount or size.

The power of compounding (exponential growth) governs many important aspects of life in this universe of ours. If a quantity reduces at a rate proportional to its value, it is then termed exponential decay. The power of compounding, this time in a negative direction, is still controlling the event.

---

[99] Currency units can differ. The specified unit of currency could be an equivalent ¥, €, £, or any other.

[100] It should be noted that inflation will usually reduce some of the benefit of the compounding effect, but it is also usual, in times of high inflation, for interest rates to likewise increase, thus the strategy remains valid over the longer term.

A difficulty with our understanding of climate change is that we are still quite low on the exponentially compounding growth curve, in relation to the absolute quantity of CO_2 in the atmosphere. Even though the increase of 43% in atmospheric CO_2 compared to pre-industrial levels is significant, it has taken more than two centuries to become sufficiently problematic to actually start melting the polar ice caps, land based ice sheets and major glaciers.

Using again the $25 per week analogy, it takes close to twenty-six years to reach just over $60,000, but less than another fourteen years to triple to the $183,000 balance. The power of compounding is relevant to the build-up of CO_2 in our atmosphere, with the added concern that in the financial example, the rate of compounding was constant for the entire forty years, but in the case of the atmosphere, the rate of build-up of CO_2 is also increasing.

Over the current one hundred years of the Industrial Era, compared to the first one hundred years, we are putting far more CO_2 into the atmosphere than was the case when James Watt invented the modern steam engine in 1765. The improvements in fossil fuel-burning efficiency over time do virtually nothing to counteract the monolithic increase in the tonnage of coal, oil and gas currently being burnt.

Where the Rubber Hits the Road

As the atmospheric CO_2 builds up and the Tundra regions and the Arctic Oceans warm, the permafrost will begin to thaw and the potential for vast quantities of methane gas to be released into the atmosphere will be unstoppable. As extra methane is added to the total greenhouse gas equation, even more methane will be released. The power of compounding yet again—making the need for action even more urgent.

As previously discussed, a multiplying effect of the melting ice includes a reduction in fresh water to some very important population areas, because as the glaciers become smaller, less headwater will be supplied to some major river systems[101]. Melting of land-based ice sheets will also raise sea levels and inundate prime coastal agricultural land.

Combined with this, as these land-based ice sheets melt, which are often covered with fresh snow, the average albedo of the planet will be lowered because this ice has a significantly higher albedo value than open sea or land,[102]. This will lead to the retention of more solar energy.

To a large extent, both land and sea based ice is covered in snow, therefore any loss of any ice will add to retreating snow lines and accelerate the reduction in overall planetary albedo even faster; speeding up even more, the warming process. The power of compounding is everywhere.

One area where there appears to be some consistency of thought is the relative stability of global temperatures for almost two thousand years leading up to the beginning of the twentieth century. However, since 1900 the temperature has been steadily increasing.

It is one thing to talk about financing adaptation to the changing climate and quite another to guarantee that the adaptation will be effective. The effects of rising temperatures will also compound. This will make adaptation programs increasingly problematic, especially in vulnerable low-lying and often less affluent countries.

An Exponential Calculation

Bearing in mind the obvious difference in the amount of CO_2 emitted in 2015, compared with the early years of the Industrial Era, a reasonable assumption is that the growth rate in greenhouse gases in general and CO_2 in particular, is exponential. Based on that assumption, atmospheric CO_2 would be growing according to an exponential equation.

---

[101] An example is the Ganges river system in the Indian Subcontinent.
[102] Albedo is discussed more fully in Chapter 8

If we were to assume that the *rate* of increase of CO_2 has been constant over the entire industrial era, we could then hypothesise that current atmospheric CO_2 would be equal to the pre-industrial CO_2 level, times the rate of increase, raised to the power of the number of years since the start of the Industrial Era. That is:

$$Current\ CO_2 = Pre\text{-}industrial\ CO_2 \times Rate\ of\ Increase^{\wedge Years}$$

Knowing that the pre-industrial level of CO_2 was about 280 ppm and the 2015 level is over 400 ppm and taking the effective start of the Industrial Era as 1765[103,] this gives a time span of 250 years. Solving for the rate of increase we get approximately 1/7 of 1 per cent per year or 0.00143 per cent.

$$Therefore,\ CO_{2\ (2015)} = 280\ ppm \times 1.00143^{\wedge 250yrs}$$

Of course, this formula will yield a value for atmospheric carbon dioxide in 2015 of approximately 400 ppm and this obviously correlates with the present concentration. This was the methodology underpinning the formula.

Considerably more complex data sets will be necessary to allow for all possible correlations between pre-industrial and current levels of atmospheric CO_2. Also, as earlier foreshadowed, it is certain that the *rate* of increase in CO_2 is also increasing.

It is not possible that atmospheric CO2 was increasing at 1/7 of 1% per year in the eighteenth and nineteenth centuries. Therefore, the exponential rate of increase over the past fifty years must be substantially higher than that of earlier years.

---

[103] James Watt invented the modern steam engine in 1765

164

Technological Compounding

When James Watt invented the modern steam engine in 1765, steam engines had been around for some years, but the efficiency was very low. The technological breakthrough credited to James Watt was the incorporation of a separate steam condenser. This invention enabled the rapid expansion of coal fired steam power to become the backbone of the Industrial Era. In later years, oil and natural gas has supplemented, but not superseded coal as the fossil fuel of choice in many applications.

If we consider the latest technology fighter jet or passenger aircraft and compare it to Watt's 1765 'separate condenser' steam engine, we see the power of technological compounding. Both machines use fossil fuels to power vastly different engineering systems.

The question is then 'how many generations or iterations, of technology are there between James Watt's steam engine and our modern jet aircraft?' They are both fossil fuelled powered machines, which are generations of development apart; an elegant example of technological compounding. This is the power of compounding at work.

Comparing the many generations of development embedded in today's fossil fuelled machinery and transport, to the minimal generational development of non-fossil fuelled alternatives that will be required in the future, it is clear there is a *huge 'generations of development' gap*.

The power of technological compounding has not been able to work its magic on zero CO_2 emission energy and fuel technology, because for the entire Industrial Age easier alternatives existed—based on fossil fuels. Accordingly, there was no requirement for scientists and engineers to develop zero CO_2 emission technology, machines or equipment.

Consequently, the technological compounding needed to accelerate the development of alternatives to fossil fuels, for base-load power generation and transport fuels, has never occurred. With a few exceptions such as hydropower and naturally occurring geothermal power (hot springs) everything else is effectively 'first generation'.

This is not to detract from the very sophisticated technology in photovoltaic and mirror-based solar thermal power generating systems, or in hydrogen fuel cell applications for transport.

It is simply that the development of technology that emits zero CO_2 into the atmosphere has not yet benefited from the power of compounding to the same extent as fossil fuel based technology.

Unfortunately, we do not have 200 years, 100 years, or even 50 years to do the 'catch-up' work. This is the challenge of our age.

This is why we must accelerate the development of the technologies of the future. This is why we must restructure our economic paradigms.

This is why we must increase resourcing of scientific and engineering research into non-fossil fuels, by orders of magnitude. Then, the power of compounding technological iterations will work to reduce emissions, stabilise atmospheric CO_2 and underwrite our collective futures.

The economic paradigm shift must start now, which will demand large capital expenditure and a great deal of research before the gains are realised. But it will be worth it.

Gradualism

Economic gradualism is a strategy that can underpin and provide the impetus to implement the power of compounding for CO_2 emission abatement. The role of governments in changing public behaviour through various economic initiatives is unarguable.

However, in modern politics, almost every government policy statement appears guaranteed to attract criticism from some quarter. A deliberate policy of easing gradually into new atmosphere-friendly economic policies may be a method of countering the collective negativity about anything designed to reduce the emission of greenhouse gases.

Aside from the negativity of very powerful lobby groups to new initiatives to counter global warming, it is also clear that the world leadership team cannot currently agree on the sweeping economic reforms that will be necessary to reduce greenhouse gas emissions. After more than twenty-five years of talking and many global conferences, if sweeping reform was going to happen, it would be global policy by now. Clearly, a different approach is needed—a new perspective.

Gradualism in Action, or Not-in-Action

The following are two examples from Australian domestic economic policies. These examples are used to illustrate the possibility of gradual, but progressive solutions to the breaking of the 'log-jam' currently preventing concerted, worldwide, economic action to reduce and ultimately eliminate, the carbon dioxide emissions into the atmosphere that originate from human activities.

The following illustrations are 'small' economics compared to the task of reducing anthropogenic CO_2 emissions, of between 35 billion tonnes and 40 billion tonnes[104] annually, to a net zero value—in a time short enough to protect the complex structures of our civilization. It is the concept that matters.

The first example is a policy designed to arrest a decline in private health insurance in Australia. The scheme imposed a premium increase of 2% for each year that a person delayed taking out private health insurance beyond their 31st birthday, capped at a maximum of 70%. Soon after the introduction of

---

[104] Including emissions from land use changes; see Chapter 3 – Complexity.

this policy, with the numbers of people having private health insurance increasing markedly, it can be concluded that the methodology was successful.

It must be said that no aspect of public policy is simple. In addition to the escalation in premium rate there was also a substantial rebate of the insurance premium through the tax system and an amnesty period that enabled older people to lock in the '30 year old' rate. At the same time, and depending on taxable income, the levy for the universal health care system was increased for taxpayers who chose not to take out private health insurance.

This example only concerned one small sector of a mid-sized economy; we can only guess at the complexity of decarbonising the global economy, but again, my view is that humans are smart and are capable of meeting the challenge.

The success of the above scheme in raising private health insurance participation rates in Australia was and is, a fine balance of 'carrot and stick' and serves to illustrate how the principle of economic gradualism can be used to change behaviour, a strategy that could be scaled to a global context.

The second example, also in Australia, shows where an opportunity to gradually implement a change in economic policy was not chosen, but instead an 'all at once' approach was adopted, which subsequently failed. Australia has a policy where losses from an investment in the property market can be used to offset personal 'wage and salary' taxation.

It is one of only a few countries in the world to allow this type of tax offset without effective caps and conditions, to limit adverse effects on the taxation system and owner occupiers of housing. The policy has the unintended consequence of forcing up the price of housing.

The policy is effectively a subsidy to owners of houses and apartments, which are bought as loss-making investments, with the expectation of reducing personal taxation and capturing future capital gains. Many have criticised this policy as being damaging for Australian families seeking to purchase a first home, a distortion of the taxation and banking system and the Australian financial market in general.

In 1985, the government changed the tax laws to only allow offsetting of property losses against other property gains, not ordinary wages and salary income. This change of policy triggered such a massive reaction in the investment community, housing rental market and in the media that the changed policy had to be reversed in a little over two years.

Instead of this 'all at once' policy, if in 1985 the proportion of interest and other expenses of a rental property that could be claimed as a deduction against wage and salary income had been reduced by a very modest amount of, for example, 2% and the deduction in this claimable amount then reduced by a further 2% each year[105], the instant massive disruption to the housing rental market would have been avoided.

By today (2015) the allowable deduction amount to personal income would have been 40%. By 2035, the market distorting policy would have been completely reversed. In effect, the benefits of the changed policy would have manifested earlier, because the signal from government would have been unequivocal and subsequent investment decisions made in accordance with the changed reality.

Whether the policy in question is good or bad public policy is not the issue. The purpose of this illustration is that if in 1985 the change in policy had been implemented using the principle of 'economic gradualism' the outcome could have been very different.

The principle of economic gradualism is a strategy that will be a very valuable addition to humanity's 'toolbox' of policy options in the fight against global warming.

---

[105] Therefore the claimable amount in the first year would be 98%, in the second year 96% and so on.

Gradualism and CO$_2$

So where do we go from here? One possibility is to apply the principle of economic gradualism to CO$_2$ emission reduction. For the sake of the argument, what if there was a global agreement to encourage all motorists to switch to smaller or electric vehicles. Suppose a modest fee, to be used for zero emission research projects, were imposed on vehicles powered by fossil fuels at time of purchase.

This fee could be based on a sliding scale correlated to the engine capacity and kerb weight of the vehicle[106]. Smaller, lighter, vehicles with an engine capacity of less than, for example, 0.8 litre (800cc) and electric vehicles[107], would attract no fee.

Suppose the fee for a vehicle with an engine capacity of 3.0 litre were to be set at 5% of the purchase price in the first year of operation of the scheme and this fee is increased by 5% each year. It would be twenty years before the price of a vehicle with an engine capacity of 3.0 litres would increase by 100%.

Consumers would begin to take this into account in their purchasing decisions. Of course, this one measure alone would be insufficient to drive global emissions down to zero. The purpose here is to open up dialogue on economic gradualism, not to define every detail of economic policy.

This exercise only starts with the automobile example. Policies based on economic gradualism could eventually be expanded across other sectors of the economy and in combination with reform of the taxation system, would begin to make a positive difference.

The most *important aspect is that gradual change can limit the impact on any given industry*, or the economy of a particular

---

[106] Engine capacity and kerb weight are two possible criteria; another criterion may be the vehicle fuel efficiency.

[107] Electric vehicles are an example only; it could also include other zero emission fuelled vehicles, now or in the future, from the fee.

country and allow the impacts to be managed. The example could have been about any other factor in the emission reduction dilemma, any other factor where the world is finding difficulty coming to an agreement.

For instance, the issue could be an agreement on how to price CO_2 emissions into the atmosphere. As a case in point, envisage a process started at the time of the Kyoto agreement. A price of $US1.00[108] per tonne of CO_2 emissions could have been imposed and gradually increased as the world economy slowly adjusted to and factored in, the reality that the atmosphere was no longer a free CO_2 dump.

If $US1.00 per tonne of CO_2 had been levied in 1992 and increased by $US1.00 per tonne every year since, the price for dumping one tonne of CO_2 into the atmosphere would now be $US23.00.

Most importantly, a signal would have been sent to everyone that the atmosphere was no longer that free CO_2 dump. Obviously, it would be preferable if world powers could decide, right now, on a realistic carbon price—whatever may be needed to stabilise the atmosphere. However, if this isn't about to happen soon, starting the process using the principle of economic gradualism in 2015 is better than not beginning at all.

Of course, other developments will overtake the need for specific interventions using economic gradualism techniques. For instance in the automotive example, the normal process of automotive research and development will make gasoline fuelled cars redundant long before a car with a 3.0 litre engine attracts a surcharge of 100% on the purchase price.

Nevertheless, this type of intervention signals an unambiguous intent to both industry and consumers which will, in large part, speed up the move to a zero CO_2 emission

---

[108] The specified unit of currency could be an equivalent ¥, €, £, or any other. The United States dollar has been selected in this and other examples because of the status it has as a reserve currency.

future, by encouraging investment into environmentally sustainable sectors.

A legitimate criticism of a gradual process to emission reduction is that it is too slow.

This is true. However, because such little progress has been made towards mandatory and enforceable emission control measures, there are currently few effective alternatives. An economic gradualism policy will help to accelerate the industrial research that may be expected as part of the normal scientific and industrial progression and initiate the process of technological compounding, thus connecting the complementary pathways of gradualism and compounding.

Deep down we all know it needs more, but if we cannot give it more right now, then anything is better than nothing. At least it is a start that will begin to change the way people think about the chemical changes being made to the atmosphere and oceans—and hopefully, put in place systems that can be expanded when the tide of public and political opinion changes.

Ideas Compounding

It is not only money, technology, temperature and changes to the chemistry of the atmosphere that will compound. Ideas also compound. A dangerous idea, currently being compounded by the climate change denialist, is that the enhanced global warming we are currently experiencing is not real, is part of some natural cycle or is nothing to be concerned about. This is not true.

As previously discussed, the denialist claims about global warming cannot be supported with recent data or qualified peer-reviewed climate research. Many people have stated: 'We are witnessing a climate specific version of the asbestos, smoking and lung cancer debates all over again'.

Currently, the power of compounding appears to be working against those of us who want to see more done to combat the continuing alterations to the chemistry of the atmosphere. We need to work to accelerate deployment of the technologies of the future and bring forward the scientific and engineering innovation, which would otherwise be expected to mature in the second half of the century, to the first quarter of the century—in other words, now!

Fifteen years of this current century and climate change mitigation opportunity, have already passed. We must engage the power of technological compounding onto the side of the Earth's biosphere.

A powerful reason for writing this book is the hope that it may stimulate discussion among people of all disciplines and persuasions about the best strategies to accelerate the move to a non-fossil fuel future. It contributes some ideas, including the possibility of entirely new scientific, engineering and economic paradigms to this discussion. The sheer scale of the mathematics, physics, engineering, chemistry, economics, public affairs and inter-disciplinary research that will be needed and will be devoted to solving problems stemming from global warming will be immense.

Will it make a difference to the political and economic paradigm if we were to view global warming through the eyes of a child? Shall we still be comfortable with our environmental decisions?

16. Sustainable Manufacture

Today's manufactured products are increasingly designed with redundancy and obsolescence taking precedence over longevity of product usefulness. It is important to recognise that this obsolescence is often not based on function, or fitness to task, but on fashion.

Modern manufacturing plants and systems are particularly efficient in reducing product cost and hence price to the consumer. However, next to no account is taken of the capital and recurrent cost of CO_2 emissions to the atmosphere, incurred as a consequence of our throwaway society. Redundancy of product, in ever-shorter time spans, is a negative for climate stability.

Chapter 9 drew attention to the fact that we exist in two parallel worlds. The natural world and the manufactured world and highlighted the lack of correspondence between these two different sides of our lives. There is a choice between what we say we value: family, friends, trees, green grass, mountains, lakes, oceans, fresh air and what we appear to value. We appear to place more value on the products of the manufactured world, which in the highly developed economies seem to take up our every waking moment. Practically everything we use or do owes its existence to science and engineering, in particular to mass production manufacturing.

It would be a stark choice indeed between the natural world and the manufactured world and although it seems at times that the manufactured world is more important to us, there is room to bring the two closer, without too much alteration to our lifestyles. The way we use manufactured products is an area where we can either choose to change our consumption patterns, or gamble on keeping consumption and manufacturing patterns unchanged, at the expense of the environment.

There may be the need for rare exceptions, such as aviation. With the aircraft industry comprising such a significant part of manufacturing and because aviation has a high profile in the global warming debate, the special role of aviation is addressed later in this chapter.

Changing entrenched patterns of consumption for manufactured goods, equipment and services will be complex and depend on four critical factors:

- whether we, singularly and collectively, are able to handle the change necessary to our entrenched and familiar patterns of consumption

- whether nations, governments and the corporate world have the political determination to manage the necessary change

- pricing CO_2 emission cost into every good and service in the world[109]

- pricing and/or taxing products, equipment and services[110] to accurately reflect the total CO_2 emission cost to the environment.

True Cost

This section may appear simplistic, or a somewhat strange notion, in the context of the last sixty years of consumer product development. However, there is little that could be worse than altering the chemical composition of the atmosphere by increasing one constituent part by 43%.

This has occurred in the relatively short time of about 100 years—the second half of the Industrial Era.

We in the developed world are excessive in our lifestyle choices. This is manifested in many different ways. For

---

[109] Chapter 13: Universal Carbon Cost.
[110] Chapter 14: Funding the Future.

example, it would make little difference to the increase in our culinary skills if we were watching our favourite cooking program on an old 'thick' Cathode Ray Tube (CRT) television, or on a 'thin' flat panel model—whether the screen is curved or not.

In Australia, over the last few years, hundreds of thousands of perfectly functioning television sets[111] have been put on the roadside for pick up by rubbish trucks. This has clearly happened all over the world.

The only way this would make economic sense in a new carbon constrained world, is if the CO_2 emissions of running a CRT model are greater than the total emission cost of manufacturing, transporting and running the flat panel television. The power saving between the two televisions can only be taken into account for the time that the new flat panel or curved screen, television fulfils the expectations of the consumer.

But currently, in today's style conscious world, throwing a perfectly good appliance into landfill doesn't have to make 'whole of product life' CO_2 emission sense; it only needs to make fashion or trend sense. A critical question is: 'Are we paying a realistic CO_2 emission price for a stylish look and increasingly high definition'? Until a universal carbon cost of everything is determined, this question is impossible to answer accurately.

Maintenance & Repair

Today, as well as televisions, we also throw out toasters, microwaves, kettles, power tools, vacuum cleaners, pumps, refrigerators and a host of other appliances, tools and equipment. Often, all they require is to be repaired and this is the challenge. Without knowing the total cost to the biosphere

---

[111] With the phasing out of analogue TV networks and the need for connection through a 'set-top-box', or propriety cable network, the underlying premise is changed, but not negated.

of a new product, how is it possible to determine the economics of purchasing new, rather than repairing the existing item?

To take one small example as a first case: a 40cm (16 inch) pedestal fan. These fans have been advertised in Australian national newspapers, on special, for $A9.90. It is very doubtful that this price includes the true CO_2 emission cost of mining, oil drilling, bulk transport, product design, manufacturing, distribution, marketing and retail outlet costs.

Repair is out of the question; if a replacement motor were available it would cost more than a complete new fan. In a rational, carbon constrained world, a new fan would probably cost much more than $A9.90 and the cost of a replacement motor would be aligned with the proportion of the original new product cost that the motor represented.

Another even simpler example is the non-stick frypan. A whole frypan is frequently discarded because the non-stick coating is no longer effective and it is cheaper to buy a new frypan than get the old one recoated. It is not uncommon for a non-stick, machined base frypan to end up in landfill. It is simply beyond belief that a new frypan, from bauxite mine to retailer, creates less CO_2 emissions than a local recoating enterprise.

A third example is the microwave oven: advertised on special, for just $A49.00. It would be difficult to find a repair shop that is willing to replace a microwave generator, in a faulty microwave for less than that price. Instead, when a microwave generator fails in a domestic situation the entire microwave often goes into landfill[112]. This includes: the cabinet, the glass door, the lining, the electronics and the glass food tray—all in perfectly good working condition. This is

---

[112] In countries with effective and comprehensively used recycling systems, the environmental cost is reduced, but not eliminated. There are still considerable pollution and energy costs to the environment, in re-melt recycling and re-manufacturing products from the re-melt recycling process.

good for business, but not good for reducing atmospheric CO_2.

A final example of a retail item that could be repaired, rather than scrapped or remelt-recycled is a broken door lock. In this case a component, with a mass of only 13 grams, failed. The mass of the replacement lock, together with packaging, was over 1 kilogram.

This is a kilogram of metal and plastic packaging that must be mined/extracted and manufactured. A repair and maintenance economy rather than a replace, remelt-recycling or scrapping of the item economy, is clearly more efficient from a greenhouse gas emission perspective.

The efficiencies are manifest. In the above examples, CO_2 emissions are saved on the extraction of raw materials, manufacture of the items/appliances and transport of both raw materials and the finished product. Multiply the environmental gain on this single product by millions of products and millions of different product lines; all of which could be repaired if the spare parts were available.

If changes are made, that lead to repairs being cheaper than throwing something away and buying a new one, it would not destroy the economy and it would not destroy us. A few years ago, we were all paying considerably more for similar fans, frypans and microwaves and that was when wages were considerably less than they are now. If policy settings are adopted that recognise CO_2 emission cost, the clear winners will be the environment and the different employment opportunities that may be generated.

Maintenance and repair will then effectively become decentralised manufacture. Similar benefits to those that flow from decentralised power generation systems and savings in 'food miles' will also start to accrue in the manufacturing environment.

The pedestal fan, frypan, microwave and door lock are four small examples; there are millions more and many have considerably larger CO_2 emission footprints. If the world *abandoned the 'throwaway' society in both the consumer and industrial markets*, the reduction in CO_2 emissions across the entire manufacturing sector and the wider economy, would be enormous. It will be a modified world; this does not mean a less happy or less convenient world.

We shall still have toasters, microwave ovens, fans and the host of other devices that we have come to enjoy.

Reframing the 'throwaway' society to a 'maintenance and repair society' is not an insignificant reform. It is not too little to make a difference. It is not 'tinkering at the edges'. It is 'root and branch' reform to the manufacturing sector that will make a difference for the better.

The Industrial Sphere

Similar situations apply on an industrial scale. Many components, tools and other equipment that are currently discarded could still be functional, if repaired or used by another company, or business. The economic foolhardiness of the current situation would be more obvious if the true cost of the total carbon emissions, needed to bring the new industrial equipment to the market, was built into both the new and the discarded item. The economics of industrial equipment management would then become very different.

Manufacturing Challenge

In addition to focussing on repair and maintenance, there is a further challenge for the world's manufacturing industry. This is the task of designing stylish products that immediately capture a consumer's attention, are able to be repaired and maintained and can also justify the cost of manufacture, distribution and retailing, in terms of CO_2 emissions.

The total CO_2 emissions associated with making and marketing a new product, plus its operating emissions over the expected life of the new product will need to be less than the operating emissions of the product it replaces, taken over the expected useful life remaining in the product being replaced.

As the life spans of new and innovative products are extended, by designs that accommodate spare parts availability and repair, the net emissions saving benchmark will become easier to achieve.

None of this is new; it is only the focus on CO_2 emission cost that is new. There are many examples where a new product range places fewer burdens on the environment because of a reduction in raw material usage.

Economy in raw material usage has underpinned manufacturing engineering for a very long time, but in the past, the effort has been more focussed on reducing the financial cost of products, rather than CO_2 emissions.

There has always been a strong correlation between miniaturisation and a reduction in the use of raw materials, which has clearly led to a reduction in CO_2 emissions, long before global warming was recognised as a major issue.

A classic example of more efficient use of resources, by orders of magnitude, is the modern smart phone compared to the first mobile telephone in a big-bag. The best example of all is probably the current technology integrated circuit, computer chip, with the processing equivalent of more than a billion thermionic valves[113]. The saving in materials and energy costs are astronomical.

'And', not 'Or'

The mixture of repair rather than redundancy and a requirement to justify the CO_2 emissions cost of producing a new product range, against the CO_2 emissions savings in running costs of the old product range, are equally important.

---

[113] From the pre-transistor age; a typical thermionic valve would have been up to 10cm high and 3cm in diameter.

Both strategies will be necessary for the manufacturing sector to contribute effectively in the battle to preserve Earth's existing climate zones that are, in general, benign to our civilization.

Aviation, Automobiles and the 'Way of Life' Test

What does the world do with the aviation sector? Aviation is an essential part of many people's lives. It is doubtful that even the most advanced scientific and engineering research will benefit the aviation sector, with some remarkable new propulsion system, in the short to medium term[114].

Development of aircraft propulsion systems that are totally independent of hydrocarbon fuels, are in a different scientific and engineering realm to, for example, the development of base load solar-thermal electricity generating plants.

It seems, at this stage of human development, we have no practical alternatives to the airplane and air travel. It is difficult to imagine our current world infrastructure operating without aircraft. After all, the point of massively increasing funding for advanced scientific and engineering research is to preserve the enjoyment of 'life as we know it', not to destroy it.

Decisions may need to be made on what is essential to our way of life and what is not. Or more realistically, 'what will change our way of life so substantially that it will no longer be the way of life that is so familiar to us?' If that test is applied to aviation, a great many people would argue that a world without air travel would not pass such a test.

Similarly, a world without automobiles would not pass the test for most people. A different consideration however, is the comparison between a large motor vehicle and a smaller vehicle. It is not uncommon to see one person, driving

---

[114] Say, between ten and thirty years.

around suburban streets, in a vehicle with a three, four, five, or even six-litre engine.

The larger vehicles can have a kerb weight of over two and sometimes up to three tonne. These vehicles have equipment to protect the occupant from excessive heat or cold, save him or her from physical exertion and can transport the person anywhere in quiet comfort, at higher speeds than is legal in most jurisdictions.

It is also not uncommon to see a single occupant, driving around suburban streets in a vehicle with an engine capacity of less than one-litre and a kerb weight of about three quarters of a tonne. This vehicle has equipment to protect the occupant from excessive heat or cold, save him or her from physical exertion and can also transport the person anywhere in quiet comfort, at higher speeds than is legal in most jurisdictions.

The added emissions cost to the atmosphere of a vehicle with a kerb weight of two and a half tonnes, compared to a vehicle with a kerb weight of less than one tonne, is not confined just to the extra fuel that is used by the heavier vehicle. It is difficult to think of any part of the manufacturing and supply chain that is not also magnified.

Larger tooling, a greater mass of specialised tool and die steels, larger presses and plastic moulding machines, increased factory floor space and mechanical handling capacity, more raw materials in the form of aluminium, steel and plastics, increased energy demand in the manufacturing processes, distribution and dealership facilities are all scaled-up in large vehicle production. The opposite applies in the case of smaller vehicles where the benefit to the environment is the only factor that is scaled-up.

It has been suggested that the overall (average) fuel economy of the Australian automobile fleet has changed little between 1968 and the present day. It is difficult to verify this assertion, but by an observation of the physical size of vehicles in the current automobile fleet, it is entirely possible.

There is a plethora of large pick-up trucks, sport utility vehicles (SUVs) and crossover vehicles. Also, larger twin-cab or single-cab vehicles have replaced the old coupe 'utes' from the 1960s.

A component of every advance in engine design and fuel economy is partially negated by an increasing percentage of larger vehicles with more and more features. This in turn affects the average economy of the entire fleet. At the supermarket recently, I saw a large SUV parked in a space next to a Mini. From the perspective of a sustainable environment, a 'fitness to task' test springs to mind.

It can be argued that a society where small cars are overwhelmingly the norm still passes the previously mentioned, 'life as we know it' test and our way of life would not be changed beyond recognition.

If a small car worldview became common, big cars would not be scrapped, as that would be the worst possible environmental outcome. There would be many commercial uses, particularly in the developing world, for the larger vehicles that would become redundant in highly developed countries, cities and communities. What would change is the mix of vehicles produced in the world's automotive manufacturing plants.

The automotive example demonstrates a graded test. That is, big cars versus small cars, as opposed to a 'cars' or 'no cars' choice. In the air travel example, the traveller does not have the choice to downsize the turbofans to save greenhouse gas emissions.

With air travel there is not a graduated scale. Still, to minimise damage to the biosphere, from high-flying aircraft, there are steps that can and are being taken. Bio-fuels may play a part. However, the competition between land for food production and land for algae farms and bio-fuel crops will almost certainly become problematic. With the world population expected to increase to over nine billion people by 2050 the situation is unlikely to be easily resolved.

Design improvements in aircraft, from a fuel economy standpoint, are an obvious environmental positive. Also, teleconferencing and conceptually in the future, even holographic conferencing if it was desired, could help minimise air travel to an extent.

Heavily discounted airfares are sometimes even below cost, where the price of flying short and intermediate distances is sometimes not much more than airport fees, government taxes and/or the taxi fare to the airport. This form of marketing may no longer be viable in a world that is working to reduce CO_2 emissions.

Existing modes of passenger transport for these shorter distances, such as rail and road, will become more competitive because they are more efficient in emission terms. This makes them more appropriate to a new economy that allocates a realistic cost of CO_2 emissions to passenger transport. More importantly, whether the mode is road or rail, it is already known that both of these methods of surface transport can use electricity as a propulsion system.

It only then remains to ensure that the power generation for this transport uses solar technology.

However, with air travel continuing to increase, this is one sector of the economy that will have to rely on massive emission reductions in other sectors to subsidise the jet aircraft emissions, at least for the foreseeable future. Even after every effort is made to minimise aircraft emissions, aviation will remain an emission cost that society will probably choose to tolerate. This will be one emission cost that must be managed if we are not to change our current way of life beyond recognition.

Space travel, for scientific research, tourism and exploration, is another area where it will be a long time before a zero emission environment can be achieved.

Consumerism and Manufacturing

The Earth is bathed in more than enough energy for all our needs. The aim must be to efficiently harvest this energy, make it storable, portable and transmissible, over intercontinental distances with negligible losses.

Until we can harvest solar energy, or develop other sources of zero emission base-load power, the demands on the manufacturing sector must be moderated. The sector will need time to move to a repair and maintenance model and also adjust to environmentally sustainable new products that are justified on a lifetime emissions accounting system.

In this transition phase, if consumers wish to acquire energy, or resource intensive products, for example very large automobiles, then in a free enterprise economy they must be allowed to do so. The only proviso must be that the environment is fully factored into the pricing of such items. This will make these products considerably more expensive and allow the price elasticity of demand to appropriately influence purchasing choices.

If such limits are imposed on our energy consumption, they should only last for as long as it takes science and engineering to catch up with our voracious appetite for materials and energy. Once the energy and materials used to produce a good or service are obtained from zero emission sources, the price should decrease accordingly.

It is important that existing and emerging centres of manufacturing are made to urgently consider the above changes to product development philosophy. There will be just as many profitable trade opportunities in the future as there are now.

These opportunities will be for enterprises engaged in manufacturing products that are sensitive to the long-term requirements of the Earth's biosphere and population. Consideration must be given to the 'whole of life product cost' including all CO_2 emissions as outlined in Chapter 13: Universal Carbon Cost.

In an emission constrained world, we should reduce the role that the consumer and consumerism play as *the major driving force* of the global economy. Consumerism can still underpin our daily lives, but it must be carefully managed, for as long as it takes to reframe society and productive capacity to a *new zero carbon emission* future.

17. A Risk Management Perspective

There is little evidence of corrective measures being taken that are commensurate with the climate risk we are facing. Management of the risk associated with global CO_2 emissions is made more complex because no single country, organisation or person is solely responsible. The case is then made, that it is pointless for any nation to act alone to reduce CO_2 emissions unless there are worldwide enforceable agreements in place and indeed, this is the current situation.

Admittedly, there may be short-term benefits for any one country to adopt this argument; however, the unwillingness for any country to be the first mover, does nothing to mitigate the risk—it will increase it. In relation to the Earth's climate, there are potentially seven billion individuals acting logically, but at the same time, essentially out of self-interest. We should all realise that the 'carbon sink' capacity of the atmosphere is a finite resource.

Everyone *should* have an abiding interest in the future. It is puzzling that we are not all more concerned about the rapidly increasing level of greenhouse gases in the atmosphere. It is a natural human instinct to protect loved ones from risk: our spouses, parents, grandparents, siblings, children, grandchildren and friends.

In practically every aspect of our lives, we seek to reduce risk: motor vehicle and airline safety measures, health and safety requirements in the workplace—the list is endless. Specifically, our cars are fitted with seat belts, anti-lock brakes, air bags and side crash protection bars to reduce the risk of death or injury in motor vehicle collisions. However, we appear to have a far higher risk tolerance for hazards that may arise from destabilising the atmosphere.

Why is this? Is it due to the lack of a perceptible and immediate threat? Is it the lack of certainty, or is it the financial cost? Are these the reasons why the enjoyment of living in today's consumer society, negate any need to take a longer-term view?

Even if we deliberately accept the risk for ourselves, is that the right decision for others?

As previously discussed, currently it appears that the unspoken assumption is that technologies already existing, or under development will be sufficient to fix the problem of global warming. Even if this is not official policy, it seems to be the default position of many people in authority throughout the world. In view of the complexity and scale of global warming and climate change, this 'steady as you are' approach is based on a manifestly over-optimistic assessment of the situation.

From a risk management perspective, believing that existing technologies will be sufficient to fix the problem is akin to having no house insurance. It is not adequate protection that just because the electrical wiring, heating system and other appliances are in good condition, there is no need for house insurance. Many people pay for insurance for a lifetime and rather than thinking it is a waste of money, they are grateful that the house did not flood or burn down.

Protection of the climate is in exactly the same category as house, motor vehicle, sickness and travel insurance.

Of course, the protection of the biosphere is substantially more complex than property and personal insurance. There is no insurer that can underwrite the policy and if we do encounter a 'worst case scenario' situation, to whom would we send the claim form? And, how would we find and travel to a suitable alternative planet?

Clearly, other ways have to be found to insure our benign biosphere and the only insurance policy against catastrophic climate change is moderation of energy use and advanced technology. Mitigation of global warming demands a more determined and visionary approach to the science and engineering of new methods of harvesting energy *and the insurance premium is funding the research.*

Hazard management

Identification of potential hazards[115] is the first step in a risk management strategy. Looking at the climate situation, there are literally thousands of potential hazards that could be triggered by rising global temperatures.

The second step in the process is risk assessment. This consists of rating the risk, calculating the likelihood of the risk and the possible loss, should the particular hazard occur. An important part of the risk assessment process is to determine the financial and other impacts of the occurrence of the hazard.

The third step in the hazard management process is risk control and this involves taking measures to minimise the likelihood of a hazard occurring and being prepared to deal with the consequences should that hazard occur. In some complex situations, a range of control measures may need to be actioned.

Methods of controlling hazards and the associated risk range from a best case scenario of eliminating the hazard and removing the risk entirely, to minimising the hazard as far as possible and then providing, for example, protective equipment or apparel. Risk reducing equipment can vary from full radiation protection to safety boots and rubber gloves.

The risk assessment procedure for the planet should also follow this process. The risk must be ranked according to the severity of an occurrence of the hazard and the likelihood and frequency that the hazard could be expected to occur. The hazards arising from chemically altering the Earth's atmosphere and oceans is mind boggling in scope. The

---

[115] Hazards can be defined as situations that expose people to injury, illness or disease.

191

severity of the impacts if the hazard occurs is potentially catastrophic and even if it occurs only once, we cannot afford to take the risk. It is a hazard that must be eliminated.

Planet-wide

The need for a rigorous system of risk management for planet-wide hazards is an aspect of the global warming debate that is not given sufficient prominence. Progress on emission reduction measures may be faster, if the methodology was properly applied in quantifying planet-wide hazards and definitively assessing risk. A consequence of the lack of a risk management plan is that there is no reliable estimate of the time we have to solve the complex problems emerging in the atmosphere and oceans.

Currently any method of rating risk is inadequate for rating risk that may result from runaway global warming. The difficulty, in applying current risk rating methodology to a planet wide hazard, is that typically the highest category for assessing a hazard is the death, or permanent disability to one or more persons.

With the possibility of multiple hazards occurring as a result of global warming and with about seven billion people on the planet at risk from runaway climate change, a higher hazard category is clearly needed. Determining the rating scale for planet-wide hazards is going to be very challenging.

Natural disasters such as earthquakes, tsunamis, volcanic eruptions and hurricanes are rated on intensity scales specific to the particular natural phenomenon. It is hoped that effective CO_2 and other greenhouse gas mitigation measures are put in place long before we need to invent a 'Richter-type' scale for climate calamities.

Climate Change – Not the Only Hazard

Climate discussion usually focuses on global temperature rise in relation to different predicted levels of CO_2 in the atmosphere. There is almost no discussion on how different planet-wide levels of CO_2 will affect human health. There are already requirements that address CO_2 exposure, which are designed to protect people's health within buildings or enclosed spaces, but nothing for exposure to projected CO_2 levels for the entire planet.

There is little thought about possible safe CO_2 exposure levels for more than 80 years of continuous exposure, which is the average human life in developed countries—all of which will be lived in the atmosphere.

It is likely that atmospheric CO_2 will reach the much canvassed 450 ppm and potentially levels up to 550 ppm are possible. It is concerning that no one really knows the final CO_2 concentration.

In a few years' time, a newborn will be likely to face CO_2 levels for the whole of their lives that we would probably not allow in their schools and kindergartens for even eight hours. This is an aspect of our industrialisation that demands more consideration.

Another aspect to global warming, which has not been prominent is the human comfort temperature range. It seems generally supported that the average global temperature rise should be limited to 2°C.

Even a seemingly small number such as a 2°C increase represents a very significant part of the human comfort range. From my experience, without heaters, fans and air conditioners, an 'inside-the-house' comfortable temperature zone spans only about ten degrees Celsius.

This obviously depends on the activity being undertaken, the humidity and also the air flow through the house. Therefore, the widely accepted 2°C equates to a massive 20% of a normal human 'inside-the-house' comfort zone.

Remember too that 2°C is rapidly becoming a best-case scenario, so effectively 20% of the human comfort range used up by the best case scenario. The actual average global temperature rise could be higher.

More & More Canaries in the 'Climate' Mine

The climate versions of the proverbial 'canary in the mine' are in trouble all over the world. The oceans are becoming more acidic, from a pH of 8.3 to a pH of 7.8. Average temperatures are increasing faster in the high northern latitudes, the tundra and the Polar Regions, than in the tropics and southern latitudes—a very dangerous sign. All over the world, the land ice, floating ice, ice caps, ice sheets and glaciers, are diminishing in area or thickness, or both. All these changes represent the current default risk rating for planet-wide hazards that are already manifest and indicate a problem of immense proportion.

In terms of the planet-wide risk of CO_2 emissions, the world community is not behaving in keeping with accepted risk management principles. This is a gamble, where the odds of winning are exceedingly long. Effective action to control risk is urgent.

18. Media, Marketing, Government and Leadership

The chemical stability of the atmosphere and oceans are two critical geophysical factors on which we shall be judged in twenty years. At the centre of the fight to stabilise this chemistry, will be members of the broad population as well as community, business and political leaders. The pursuit of successful climate change mitigation measures should engage us all, both personally and professionally, to the highest level of endeavour.

First place in the position description of anyone seeking employment ought to be engagement in developing methodologies for establishing a carbon-free economy, while at the same time maintaining individual freedoms, happiness and enjoyment of life. Every individual—people at all levels of the human organisational hierarchy—will be pivotal in the battle to preserve the climate.

At the highest level of leadership are Prime Ministers, Presidents, Kings, Queens, Heads of Religion and the Secretary General of the United Nations. Next and just as important, are the people who run the world's companies and corporations, government agencies and the military. This level of leadership has the power, if they choose to exercise it, to manage their organisations for the survival of civilization as a whole and not just the particular bank, mining company, government agency, or industrial empire they represent.

Finally, the media, which is possibly the most important sector of all because of the key role it plays in the formation of public policy.

Media Power

If people within the media were convinced of the dangers of a warming planet, the political battle would be all but won. Print, radio, television and Internet journalists have far more political power than the average man or woman in the street. They are media trained, articulate and most importantly they control what is discussed, the course of the conversation and when it ends.

The purpose of media, in no particular order, is to inform, to sell advertising and to entertain. The reality is that it also has a powerful influence on the political process.

The dominant political discourse can be transformed either by direct media influence on the political hierarchy, or indirectly through the influence the media has on various audiences through print, television, radio, the internet and social media. As such, the media and marketing sector will be pivotal in getting the sustainability message into communities worldwide.

New Media – New trends

Two relatively new media trends compound the difficulty in winning the battle to preserve our climate. These are the 24-hour news cycle and the adverse consequences that arise from 'even-handedness' in media policy.

Neither trend was apparent and was certainly not a problem for Generals Eisenhower, Montgomery and their senior military commanders when planning the Normandy landings in 1944. If the Normandy landings had been subjected to the same level of media scrutiny and partiality, as is the case with the suggestion of effective action on climate change, it is doubtful that a single Allied ship would have crossed the English Channel.

Even-handed, politically correct, media policy can lead to a distortion in the way climate science is reported. A typical and frequent example of this type of media reporting occurs when a well-credentialed scientist or economist engages in a discussion on some aspect of climate change, within their particular area of expertise.

The subject of the discussion may, for example, be the emergence of unusual weather patterns, carbon pricing schemes, nuclear power or ocean acidification. The discussion often concludes with a comment by the media host such as, 'well, in the end we shall all have to make up our own minds on these issues'—as if we were buying a chocolate bar.

This is sensitive to all viewpoints and a great way of running a radio or television station, but there is a more serious imperative. Scientific decision making is far too important to be left to unqualified individuals, rather than specialists with expertise in the particular field. To make a medical analogy, when it comes to our own health, we invariably seek expert specialist opinion; the health of the biosphere demands no less.

The 'even-handed' policies also result in a relatively few, high profile, climate change denialists being accorded approximately the same media time as well-credentialed scientists, with real and rigorously peer-reviewed research on the warming planet. While hard to determine an exact number, it must now be beyond all reasonable doubt that there are many more credibly qualified scientists holding the view that the planet is warming, due to human activity, than the opposite.

The policy of according equal media space to each viewpoint, irrespective of the number of scientists holding that view, is not helpful and achieves the exact opposite of a balanced debate. Adding to the distortion, is the fact that a vast majority of those in the denialist faction have qualifications in scientific fields other than in climate science, or the chemistry of the atmosphere.

The answer to the question of whether and when to take effective action to mitigate global warming should be decided in both the affirmative and the immediate—that is what the science is saying. The 'devil's advocate' position is simply not appropriate and the media should reflect this reality. In times past, when the whole world was at war, no nation suggested giving equal media time to the opposing view. Realigning the balance in media reporting won't fix the climate problem immediately, but it is an essential first step. This is a critical issue.

24/7

The 24-hour, seven days a week news cycle has spawned a troublesome offspring, the 24-hour opinion cycle. There have recently been many commentators making the valid point that the 24-hour opinion cycle is not helpful to the cause of any long-term reform. With the focus usually being on short-term point-scoring, determined and meaningful discussions are often hampered. An additional difficulty in getting serious debate on substantial issues of policy reform, such as fossil fuel emissions reduction, is for the reason that there will always be special interests and everyone wants to win.

Consequently, the negative effect of the 24-hour media cycle is that policy debate is frequently distilled down to short slogan-like media statements; this applies to most areas of human enterprise. Action to combat global warming is an early and ongoing casualty of this phenomenon.

In economic terms, effective global warming mitigation will have a substantial cost and however that cost is structured, some commentators will almost certainly label it a tax—whether it actually is, or not.

In historical terms, climate change science is a relatively new area of scientific activity. Therefore, action to combat global warming and climate change is particularly vulnerable to the power of negative campaigning conducted in the media and magnified by 24-hour saturation coverage of the negative viewpoint. Opinion has always been a 24/7 activity, but at the corner shop or in the lunchroom, the power of opinions voiced in those locations is insignificant compared to the power of the national and international media.

Healthcare, education, defence, immigration and other critically important 'big ticket' items have been with us for years and are already being funded. While, obviously all of these areas could use extra funding, a five year delay in securing that extra funding is not going to be as critical as a similar delay in securing appropriate funds for real CO_2 emission reduction. The imperative to mitigate anthropogenic global warming and climate change is currently more urgent than these other important funding requirements.

Social Media

How the social media phenomenon influences the global warming debate is yet to be determined. In particular, the speed and flexibility of mobile-based technologies, linking the world community via the Internet, has almost unlimited potential to change the debate. Social media is an interactive means of communication with a large part of the content being defined by the user.

Whether this will benefit those people wanting action to reduce global warming or benefit those with a denialist perspective remains to be seen. The outcome will significantly depend on the sources that participants in social media forums use to obtain information.

If the users of social media platforms reflect information obtained from mainstream web-based media, then mainstream media will continue to control the debate.

Emergencies and Media

After a natural disaster, there are often discussions conducted in the media (for example on talk-back radio) on the merits or otherwise, of knowingly choosing to live in areas where there is a high probability of natural disaster. The consensus in these discussions is generally that taking into consideration the likelihood of catastrophes, such as bushfires, floods, hurricanes and earthquakes is not an appropriate criterion for selecting a place to live. Taking such factors into consideration is deemed to be alarmist.

In our current society, unless in the face of immediate and direct danger, it seems that cautious talk is considered negative and must be avoided at all cost. While this may keep the economic 'engine' on track, in the face of real danger to the whole of humanity, prudent risk assessment should not be considered negative talk—it is sensible talk!

Of course, for economic reasons, many people have no real choice in where they live. Indeed, the entire human race has nowhere else to live. This is the reality of our existence.

The International Space Station
can never be home to over seven billion people!

In early 2011, the State of Queensland, Australia, endured two significant natural disasters. A commonly heard phrase in the media at the time was, 'prepare for the worst and hope for the best.' Consensus appears to be that this is a prudent and sensible policy. So, if 'prepare for the worst and hope for the best' works on the micro level, why is it not applied on a macro level?

What is it about global warming? What is the psychological hang-up that stops us preparing for the worst and hoping for the best in relation to the planet's biosphere? National and international debate is needed on these important issues and the only facilitator of such a debate is the media. In our current culture, the global media: traditional, Internet and social networking are the new 'town criers'.

During emergency situations, the coverage by newspaper, radio, television and Internet based services shifts to saturation coverage, which is right and proper. Features of media coverage during emergencies are first and foremost official warnings, situation reports, plus specific, general and background information.

Human-interest stories regarding challenges, persistence and triumphs over adversity also play an important part in the comprehensive coverage. All of these functions need to be carefully balanced. The warnings must be timely and detailed enough to prevent loss of life, but at the same time avoid creating panic and unduly disrupting essential services.

Where does this put the media in terms of the emerging crisis in the atmosphere? As more climate originating emergencies occur around the world the media will, without doubt, perform their role with the usual professionalism. However, we are currently in a situation where it is time for prevention, rather than disaster management.

The world's media should *fully engage in the task of preventing climate induced disaster*, rather than just doing a good job of reporting it.

Exercising Media Power

Media power is exercised in many ways. In the case of radio 'talk-back', the end of a discussion often includes a partisan conclusion by the media host. A significant proportion of the media seems to be either agnostic, or openly against any action to mitigate global warming.

The position sometimes appears to be based on an ideological view that the cost of any global warming mitigation measures will in some way constitute a new tax. As a consequence of this, public opinion has tended to be swayed to the denialist side of the argument.

In the case of newspapers, the largest headlines are sometimes reserved for climate change denying articles. It is a dangerous trend, when the opinion of a particular journalist is elevated to become the news, rather than the facts. Making the opinion the news seems to be particularly prevalent in the case of articles denying anthropogenic global warming.

The media is not a homogeneous single entity and with the advent of the phenomenon of 'social media', is even more fragmented. It is made up of men and women from all walks of life, some with children and grandchildren and some without.

They are the same as everyone else, except with extraordinary opportunity to influence the political debate. While their power to influence is far greater than the average person in the street, this does not necessarily equate to either a greater interest in, or command of climate science.

This power needs to be exercised wisely as this issue affects us all. Denialist journalists and media personalities would surely have the same hopes and dreams for the future as everyone else and for the people they are close to. With rising temperatures and melting ice, strongly correlating with increased levels of atmospheric CO_2, the important question is whether a denialist media is representing everyone's best interest, including their own.

The Difficulty of Major Reform

It is becoming more and more difficult for governments to start a conversation about any agenda for major reform. Today, political leaders have good reason to be wary of advancing hypotheses, forecasts, or other statements that suggest imaginative and different solutions to any problem. Any statement by a political leader is immediately seized upon as a firm policy announcement.

Very quickly, a well-financed lobby group is established to organise a negative campaign on the issue, including in some cases, a campaign to destabilise the political leadership. The cash for such campaigns, funded by powerful self-interest groups, appears unlimited.

It is risky for a politician to take a forward thinking analytical position on a matter of public policy, even though the issue may be crucial in safeguarding the future stability of the country. The reason being is that any scoping of a possible change in policy, away from well-established norms, is immediately taken as a 'set-in-stone' controversial policy decision rather than a basis for further discussion.

This limits our ability to have real debate about major reform and is yet another reason why achieving effective action on global warming is exceedingly difficult. This is occurring at a time when nudging the political debate towards a consensus on halting global warming must be our number one priority. This will require real maturity in our political discussion.

Today's Leadership Gets the Big Gig

Selling the momentous changes necessary, in the time available, will require leadership equal to the most inspiring ever seen. The challenge of stopping atmospheric CO_2 from increasing further is of greater magnitude than putting a man on the Moon.

The demands of such a challenge require boldness in the order of that shown by President John F Kennedy[116], when he famously stated that: '... the United States should commit itself to achieving the goal, before this decade is out, of landing a man on the Moon and returning him safely to the Earth.'

It is a fair bet that a man would not have landed on the Moon on 20th July 1969, had it not been for the impetus of President Kennedy's speech on 25th May 1961. At the time, the Apollo project was arguably the supreme technological achievement in the whole of human history. Stopping further increases in atmospheric CO_2 is no less a challenge.

A goal to completely eliminate the use of fossil fuels by 2035, within 20 years, is a task on a par with putting a man on

---

[116] Papers of John F Kennedy. Presidential Papers. President's Office Files. Special message to Congress on urgent national needs, 25 May 1961.

the Moon within 8 years. Moreover, it is worth remembering that the lunar project started from a 1961 technology base. The total elimination of fossil fuels would start from a 2015 technological base. A declaration to make fossil fuels redundant, within two decades, is something we could all celebrate.

Of course there is a difference between the emerging atmospheric crisis and a race to the Moon. Competition to land a man on the Moon was a technological feat driven by the politics of the Cold War, with superpower prestige at stake. However, voyaging to the Moon did not represent an emergency or a crisis for humanity, whereas global warming does.

The multifaceted and complex problems associated with keeping global temperature rise below 2° Celsius, is first and foremost a matter of national and international politics. Among the many unresolved issues are:

- At present, there are many proposals, but so far, no concrete agreement on the cutting of aggregate greenhouse gas emissions by specified percentages and within specified time-lines. If agreement is reached on any of the targets canvassed in the media, it is still doubtful that any of these are sufficiently ambitious. We should be aiming to halve emissions before 2025 and completely decarbonise the energy sector of society by 2035/40[117]

- How to choose a base-year to calculate percentage reductions in greenhouse gas emissions that will be agreeable to all countries. The years 1990, 2000, 2005 and 2010 are variously used and/or suggested by different countries in differing emission reduction scenarios

---

[117] The complex relationship of the aviation industry and GHG emissions is discussed in Chapter 16 - Sustainable Manufacture

- How to resolve the differing expectations of fully industrialised countries, emerging economies and underdeveloped economies. How to accommodate the hope of higher living standards in developing nations, expand the world's industrial infrastructure to meet this increased demand and still meet stringent emission control targets

- Obtaining agreement on intermediate targets. Currently, there is no collective agreement on any target, including mid-term targets. Without definite and internationally agreed reduction targets, for as early as 2020, the lack of time will make it increasingly difficult to achieve a 2035 target. Emission reductions must be 'locked-in' during the early years of the compounding process

- How to allow for the new reality in which an increasingly large percentage of emissions from emerging economies are the result of producing of goods that are sold in fully industrialised nations. The truth is some of the industrialised world's emissions are effectively being outsourced to emerging and developing economies.

Comparing like-for-like, the present level of government response to the emerging climate crisis is not proportionate to government responses to past crises. The invasion of Poland, the attack on Pearl Harbour, the early lead by the USSR in the Space Race and the 2008 and ongoing financial crisis; all led to very significant government fiscal involvement and clearly stated directives.

It seems sensible that a 43% increase in one of the chemical constituents of the atmosphere should elicit similar and urgent responses.

19. Food, Water, Refugees and War

Food, water, refugees and war; these are some of the most critical issues in the global warming debate and deserve specific mention. Those of us in the more affluent parts of the world are getting away with expensive excesses in terms of CO_2 emissions.

Any solution to global warming, that is going to be ethically sustainable over the long term, cannot rely forever on the continuing meagre use of the Earth's resources by the poorest 60% of the world's population, to underwrite the CO_2 emissions of the wealthiest 40%.

Moreover, the contribution of some citizens of the world, such as subsistence farmers in under-developed countries, to the rising levels of greenhouse gases in the atmosphere has been marginal. It does not seem fair that climate change induced droughts or inundation are often more prevalent in exactly those areas of the world.

For these reasons alone, any solution to the emerging atmospheric crisis ought to be established on the basis that ultimately everyone has equal access to the available energy. If the concept that everyone has equal access to energy is adopted, or more precisely, that everyone has a right to an equal share of the atmosphere as a carbon sink, then it is unlikely that 'business as usual solutions' will be acceptable.

The challenges of climate change will be experienced across the whole world population, but will be most evident at the extremes. At one extreme are communities that rely on subsistence farming and at the other end of the spectrum are residents of mega-cities.

Mega-cities

Mega-cities are a product of the industrial world's efficient production technologies. The citizens of a mega-city will never have a choice of going 'back to nature', using decentralised power, or other ways of avoiding centralised supply chains.

In the world's cities and particularly mega-cities, we are almost totally reliant on centralised power stations and fuel depots for power and fuel, as well as on the huge trucks that supply our supermarkets every day; so essential to our modern survival.

Therefore, unless the future of millions is to be compromised, we must solve the scientific and engineering challenges of mass-scale, sustainable, non-polluting, transmittable energy and fuel.

Once again, it becomes obvious that *expediting the technologies of the future* is the only way to solve the problems of a planet where current industrial and resource management practices are increasingly moving us out of the climatic 'sweet spot', which has been the norm for thousands of years and been so favourable to our civilizations. As discussed previously, solutions will not be mono-focussed.

Of course, advanced technological solutions will be pivotal, but energy efficiency, conservation of resources, recycling and any other strategy for more sustainable lifestyles will also be important.

It is also difficult to imagine any lasting solution to global warming without massive and comprehensive use of solar energy. Advanced solar solutions will be particularly well suited to decentralised power generation in under-developed areas of the world, as well as having a central role in providing base-load power in hyper-developed mega-cities.

Food

The capital and recurrent cost of CO_2 emissions and other greenhouse gases are relevant to food choices. In the developed world, there are many of us who over-indulge in the undeniable pleasure of food and this has a greenhouse gas emission cost. As discussed in Chapter 13, 'food miles' are not the only consideration in assessing the CO_2 emission cost of food consumed, wasted and thrown away.

The total emission cost of food includes those emissions related to crops, livestock, harvesting, slaughtering and processed food production, as well as transport at every stage. The total CO_2 emission cost to the atmosphere, of the food consumed in affluent countries, must include a consideration of emissions that arise from medical interventions that are a consequence of some of our lifestyle and food choices.

Water

This is where global warming and climate change will cause severe disruption in several different ways. Apart from rising sea levels, higher storm surges and the inundation of low lying coastal land, there are people in countries bordering the Himalayas that rely on the summer glacier melt for their water supply.

In many areas of the world, including sub-Saharan Africa, populations are already experiencing long and severe droughts. Changes to weather patterns around the world were discussed at some length in Chapter 5.

The disturbing part of changes to weather patterns that are occurring as a result of the changing climate is that the poorer the country, the more difficult is the adjustment. In particular, those living in highly populated, low-lying river deltas, or islands will have most to gain by the stabilisation of the chemical composition of the atmosphere. If rapid progress is not made towards a zero emission economy, the future looks very bleak for people living in low-lying coastal areas.

Low-lying countries with considerable financial assets will be able to build sea barriers and levees, to protect against rising sea levels and more importantly, storm surges. However, those who are poor and living in countries with meagre financial resources, will be particularly hard hit, as they do not have the resource base to re-engineer their physical environment.

War

The humanitarian cost of solving political differences by the use of military force is not the only cost of war. Consideration should be given as to whether the atmosphere can afford the CO_2 emissions that result from the use of war, as a last resource method of dispute resolution.

This is not to infer that it would be possible for all military action to be eliminated—desirable though this may be. The cost of military action should be counted, not only in financial terms and the tragic loss of human life, but also in terms of the CO_2 that is emitted into the atmosphere. The total CO_2 emissions of all military action, including equipment lost and munitions expended, in just the last twenty years must be quite significant.

Refugees

The ramifications of rising sea levels, receding glaciers and more severe droughts, in some of the most underprivileged parts of the world, are obvious and have a potential to greatly swell refugee numbers. In a worst case scenario, if total inundation of some highly populated coastal areas is experienced, the number of climate refugees will increase at an alarming rate and make current refugee numbers appear insignificant.

There would be several complete books in covering the topics of 'food, water, and refugees' and it is not possible here to comprehensively cover this broad range of issues. Each of these areas will be impacted, most likely in a negative way, by global warming. If the world moves speedily towards a zero emission world economy by 2035/40 it will benefit everyone from subsistence farmers in drought-stricken sub-Saharan Africa, to 'at risk' people residing on low-lying land in many other parts of the world.

The notion of bringing the science and engineering of the future, into the present, is essential for both developed and lessor developed nations.

20. A Spiritual Dimension

As a theological gnat, I am hesitant to draw from religion; there are many better qualified to do so. However, while chemistry, physics and the whole scientific spectrum, including engineering, economics and politics necessarily dominate the content of any book on global warming, it seems appropriate to include one chapter written from a spiritual perspective. More than half the world's population follow a religion and all will be affected by global warming.

I attend church regularly and acknowledge that my belief regime will probably vary in some aspects from the beliefs of others. So, if indeed there is an intangible dimension to life on Earth and in the Universe, it would seem to follow that there may be more to this life of ours, than the externally obvious. We do know that regardless of belief, Earth is very special. Our responsibility of stewardship to the Earth and all of Earth's diverse life is paramount.

On the matter of stewardship, the Bible includes the parable of 'the Rich Fool', which addresses the acquisition of more and more material possessions. This exemplifies a philosophy that has prescience for our current climate dilemma. This parable embodies thinking that could guide our societal values, if we are going to rein in the runaway greenhouse gases in the atmosphere. The passage in question is from the Gospel according to Saint Luke:

Then he (Jesus) said to the crowd, 'Take care to guard against all greed, for though one may be rich, one's life does not consist of possessions.' Then he told them a parable;

'There was a rich man whose land produced a bountiful harvest.'
He asked himself, 'What shall I do, for I do not have space to
store my harvest? And he said, 'This is what I shall do: I shall
tear down my barns and build larger ones.

*There I shall store all my grain and other goods and I shall say
to myself, Now as for you, you have so many good things stored
up for many years, rest, eat, drink, be merry!'*

*But God said to him, 'You fool, this night your life will be
demanded of you; and the things you have prepared, to whom
will they belong? Thus will it be for the one who stores up
treasure for himself, but is not rich in what matters to God.'*

There is a truth and a lesson in this parable, for all of us in
the 'rich' parts of the world.

Stewardship

Whether a person believes in a God or whether a person
does not, from the perspective of stewardship of the Earth
and every living creature on its surface, mankind's role may
not differ too greatly. Looking after the biosphere is either our
sacred duty, or our sensible responsibility and we neglect it at
our great peril.

Statements such as: 'Be fruitful and increase in number; fill
the earth and subdue it. Rule over the fish in the sea and the
birds in the sky and over every living creature that moves on
the ground,' as recorded in the book of Genesis, are
sometimes argued to mean that humans have the freedom to
do whatever they like with the natural world.

Expressions such as 'having dominion over' also seem to
encourage a freedom to do as we please. I would argue that
'be fruitful and increase in number' may have been applicable
to a world population of less than ten million, but not for a
world population of over seven billion. Similarly, dominion,
rule and subdue could be interpreted as the need to be
efficient custodians and managers, rather than to encourage
waste, carelessness or exploitation.

From a theist point of view, one could say that God gave
us the Sun, as part of the elegant creation, safely placed
150 million kilometres away from Earth, just right for all of

our ongoing energy needs. One could ask whether we are making the 'highest and best use'[118] of the extraordinary amount of solar energy that the Earth is bathed in each and every day.

It is impossible for any of us to understand the mind of God, but was it ever intended that we should be burning 'stored sunlight', in the form of carbon-based fossil fuels, for more than a very limited part of our history? Maybe, we should consider fossil fuels as no more than an interim solution. Of course, a similar conclusion may well be reached from a more agnostic point of view.

Surely, looking after the atmosphere is part of the responsibility of stewardship. The stewardship role of mankind and our relationship with the environment is not going to be too different, whether God created the universe including the atoms that eventually made up all of us, or whether the 'Big Bang', nearly fourteen billion years ago was a result of an event, the exact trigger for which physicists and cosmologists are still researching.

The day of reckoning may not be some apocalyptic event as some have postulated throughout history. If mankind does not quickly solve the CO_2 emission problem, the decline of civilization may occur over many years and we may have more than enough time to contemplate our inadequate stewardship and our part in the destruction of this Eden.

The Unexplained Universe

As we attempt to explain every detail of the science, operation and purpose of the universe, from the smallest sub-atomic particle to the very edge of the expanding universe, there appears to be more unknown than is known. Heisenberg's uncertainty principle is a case in point, where the act of observation or measurement, is said to change both the

---

[118] A real estate industry term, pertinent to land utilisation, regarding whether the 'highest and best use' is being made of that land.

observed and the observer. Also, a unified field theory, to bring together the science relating to the behaviour of elemental particles and the physics of the macro-universe, seems to be as elusive as ever.

It was apparently said by Voltaire in 1770 that 'If God did not exist, it would be necessary to invent him[119]'. This quote has us thinking of the concept of sub-atomic particles that some have called the 'God-particle'. It sometimes seems that, at the innermost and outermost limits, our highest science runs out of explanations for every detail of the universe.

The beauty of a tree; a snowflake, or distant nebulae—all of these things are impossible for humans to reproduce. How much of the universe and our role within it do we really understand? There is such a tiny, tiny part of the universe we occupy and influence, making the concept of 'God' as reasonable a hypothesis as any other, to explain the magnificent unknown.

If monastic life is at one end of the spectrum, where is the other? Does it really matter whether God is a product of innate faith, or whether the explanation for the Universe will eventually be discovered by humans through science? Currently, there are almost as many unanswered, as answered, questions?

We all have our own beliefs, the exact detail of which varies markedly. The variance in the detail is unimportant; we still only have the one planet and the important thing, is how we look after it.

---

[119] Voltaire, in a letter to Prince Frederick William of Prussia (1770)

21. The Future — Ready or Not

Our civilization has always relied upon a continuing 'best case scenario'; a scenario that assumes there are no rogue asteroids, no 'super volcanic' eruptions and that any emerging 'super-bugs', 'super-viruses' and our political differences can be kept under adequate control. The only reasonable, optimistic and prudent standpoint is one that assumes that none of the above scenarios trigger a catastrophe.

On occasions, however bizarre, it is heard that a justification for inaction on global warming is that the Earth may be hit by a large asteroid, which would destroy our civilization anyway. This is just another small part of the general negativity towards taking decisive action to preserve the biosphere.

I heard a comment recently that suggested that, in some way, the present generation should be envious of those who lived during the First and Second World Wars. The reasoning was that people living during the two world wars had a clear mission in life; something that is lacking today. Apart from not endorsing a view that there should be nostalgia for war, the truth is that present generations do have a clear mission.

Arresting the warming of the planet is the clear mission for us today. The question remaining is: 'Are we up to the task, as were previous generations?'

As I said in the introduction, this book is one of optimism. It expresses the view that with political determination and substantial global action, we have every chance of success in the fight to overcome the challenges posed by the ever-increasing concentration of atmospheric greenhouse gases, stabilising them and ultimately reducing the amount of these gases in the atmosphere.

However, the significant challenges we now face demand much more than optimism, determination, and hope. It will demand an end to the current economic paradigm that is based upon a continued use of fossil fuels to underwrite our lives.

Let us all hope we are ready and up to the task before us. The effort that present generations must make to minimise the dangerous climate situation we are bequeathing to our future selves and future generations, will be immense.

Arresting further damage to the Earth's atmosphere/biosphere is a mammoth task. It will take the combined and total effort of the entire world to carry out that task.

All this work must be done without condemning the world to a cheerless, bleak, unhappy future; a world that would be anathema to us all.

Above all, the preservation of the biosphere must also recognise a need and indeed a moral right, of less developed nations to elevate the living standards of their people to that of more developed nations. The elevation of these living standards will clearly bring about a need to increase both productive capacity and energy output; the provision of which must fully factor in the capital and recurrent carbon cost to the atmosphere.

As well, the political reality is that this must also be achieved without reducing the living standards of people in the developed nations. Development of innovative technologies and fresh ways of thinking will be the only way to make this a reality.

The aim of higher living standards for everyone should not be abandoned on the grounds that it will be too difficult to achieve. In truth, a world society that is equitable for all its citizens, will be a society that goes a long way to fixing the problem of excessive greenhouse gas emissions. The planet Earth is a place of abundant resource.

The 'It's All Under Control' Illusion

There is plenty of high-level research being undertaken in every field of science and engineering ranging from: improving current industrial technologies, relativistic physics, quantum electrodynamics, nanotechnology and a host of other disciplines—far too many to list.

The broad spectrum incorporates: physics, cosmology, the biological sciences, chemistry, civil, chemical and mechanical engineering, mathematics and economics—again to name just a few. There is little doubt that given time, the world will no longer rely on fossil-fuelled energy systems. The problem is *this is not happening fast enough.*

The challenge of our Age is that in the time it will take for new scientific and engineering discoveries to replace the fossil-fuelled methods of doing business, the biosphere may have become quite unfriendly to our neatly ordered society.

The question of our Age is; are we prepared to embrace the magnitude of change needed to enable technologies to be developed and deployed by 2030, which otherwise are unlikely to be developed until much later in this century?

We live with change all the time and an unfortunate side effect is that the word itself has become devalued. The changes we must now embrace go far beyond the thickness of our televisions, the elegance of our latest kettle, or the styling and method of operation of a million other consumer goods. The challenge we face is to massively reduce the world's energy intensity, this being the units of energy required to produce each unit of Gross World Product (GWP).

Any reduction in energy intensity applies not simply to a comparison between the capital energy costs of one product, service, item of infrastructure, plant or equipment and another, but should also include the emission costs of maintenance over the life of the item and its eventual replacement. This is applicable to all products, services and capital goods that contribute to the emission of any of the greenhouse gases.

As humans become more adept in utilising the boundless solar energy, it will be less important to reduce the energy intensity per unit of GWP. This is the future that we work towards—it is not our current situation.

> *The constant and inescapable reality is that the necessary high-level, climate change mitigation scientific and engineering research will not be cheap, but it will be essential for our continuing quality of life. It is an illusion that current private and government funding is sufficient to achieve 'fossil fuel free' energy independence, in a timeframe sufficiently short to be truly effective.*

A Process

Significant reduction in atmospheric CO_2 and other greenhouse gases will not happen if we continue with fossil fuels, the industrial use of which dates back more than 250 years. While our existing technologies for low CO_2 emission sources of energy and fuels, such as wind, geothermal, hydropower, tidal, bio-fuels, nuclear, photovoltaic and first generation solar thermal plants, are essential in the short term, they will not be sufficient in and of themselves.

New methods of harvesting and transmitting 'base load' solar energy, on a massive scale, must be developed to replace existing fossil fuelled energy sources to supply the mining, extractive resource drilling, manufacturing, transport, commercial, retail, agricultural and domestic requirements for continually expanding markets.

Many iterations of technological compounding will be required to replace all the coal, oil and gas being used in 2015, without even considering future requirements. The world must move from this coal, oil and gas dependency for its electricity generation and the massive energy requirements of millions of energy-hungry industrial processes. Current business as usual energy and fuel management strategies are not able to supply these needs without recourse to CO_2 emission-intensive fossil fuels.

It is critical that methods are found to accelerate developments in science, engineering and methods of manufacture vital to pull humanity out of the ever-deepening hole we are digging for ourselves. The appearance that global warming is not yet creating an immediate emergency or crisis; or posing a direct threat, or having a material effect on economic activity, is not an excuse for inaction.

In a system as large and complex as the biosphere of the Earth, critical tipping points could be passed that we shall not notice in our daily lives, until much later and maybe even too late. Remember the paradox of the Phoney War. Not being able to see the carbon dioxide in the air we breathe, does not make its increase in the atmosphere less real.

There are no grounds for minimalist action on global warming.

The Road Ahead

Although we know why we should rapidly phase out fossil fuels, we do not yet know how we can do this on a global industrial scale.

Recalling how the power of compounding works; in the first ten years or so, not much happens. The benefits only start accelerating after twenty, thirty and forty years. In the case of fossil-fuels technological compounding, the world has seen 250 years of it, with about 99% of the fuel and energy effort concentrated on developing technologies for more efficient fossil fuel burning machines, rather than the development of non-fossil fuel systems.

If we had started in the early 1990's, there may have been the option of slowly changing the way we do business. Twenty years on, the softly, softly options are rapidly disappearing and we must now speed up the transition to a low-emission economy. The alternative is to let the physics of the atmosphere/biosphere impose unknown and maybe even more rapid changes on us.

When the world starts to seriously develop the non-fossil fuel technologies essential to stabilise the biosphere, having even forty years for the power of technological compounding to work on them, is looking unlikely. So, in 2015, we cannot afford to lose a single year before really getting started.

This is the very heart of the matter. Because any progress arising from compounding is hardly noticeable in the first ten years, there is always a temptation not to start. This is a devastating trap. The power of compounding cannot work, unless the process is actually started.

It is essential that everyone is aware that even if the highest priority is placed upon reducing greenhouse gas emissions right now, there will be virtually no stabilisation of the chemical composition of the atmosphere within the next few years and maybe even decades.

That is the reason the campaign must be started without delay. That is why in 2015 everything must be thrown at this task.

Everything

So, what is everything? The roadmap to a fossil fuel free future, as outlined throughout this book, can be distilled into the following themes.

It is engaging the global citizenry as well as political, religious and business leaders.

It is developing economic policies to overcome community and business resistance to the implementation of significant greenhouse gas emission reduction measures. With millions of conflicting interests, economic gradualism policies, on a large scale, may be the most practical model to underpin the de-carbonisation of our consumer society. Economic gradualism is a far-reaching concept not in any way comparable to conventional business as usual models.

It is massive funding of scientific, engineering and technological research and development into advanced climate change mitigation. This will compress the time required for successive iterations of technological compounding and pull forward the non-carbon energy systems of the latter half of this century to 2030—or earlier.

It is according the highest possible priority to relegating fossil fuels to a mere bit player in the energy and fuel cycle. The first achievements should be rendering obsolete the need for combustion of fossil fuels in all land and sea based operations, with aerospace operations following as soon as possible thereafter.

It is finding new methods of short and long-term energy storage to provide for mobile applications and transport.

It is the considerable challenge of finding methods of transmitting energy over long distances, with negligible losses.

It is taking advantage of all currently available zero emission energy, power and fuel sources: wind, solar, tidal, geothermal and nuclear fission. Decentralised power generation and the avoidance of extensive transmission grids, will also play a key role.

It is soil conservation, reforestation, water, energy, species diversity and bio-sequestration. If the technologies can be proven and are economically viable, it is also carbon capture and storage (geo-sequestration). Depending on suitable technologies, it may even include geo-engineering[120].

---

[120] Geo-engineering is a proposal to cool the planet by scientific and engineering means and has been widely criticised as being too dangerous. The potential danger would arise from modifying the atmosphere and the oceans without necessarily knowing the full effect. However, what is not widely recognised is that humans have been engaged in dangerous geo-engineering for over 250 years - by increasing the level of atmospheric CO_2 from 280 ppm to over 400 ppm since the dawn of the Industrial Era.

It is ensuring that repair and maintenance, non-destructive recycling and re-manufacturing become more important than obsolescence, redundancy, energy-intensive remelt-recycling and dumping into landfill.

It is keeping the focus on all currently available energy conservation measures. Appliances switched off at the power outlet (instead of standby mode), high efficiency (low wattage) sources of illumination and a massive increase in automotive fuel economy standards, public transport wherever feasible, car-pooling—the list goes on.

Old Endings and New Beginnings

We must think of global warming as a threat to our well-being as serious as WWII and consider the required financial commitment from this perspective. As such, a campaign to commit 8% of Gross World Product to atmosphere stabilising scientific and engineering research and development should begin immediately.

The entire rationale for such well-funded, high level, scientific and engineering research is to provide the clever solutions necessary to achieve full deployment of solar and other zero emission technologies and eliminate greenhouse gas emissions from all sectors of the economy.

People will still want to enjoy a decent life and that will demand prodigious amounts of energy and fuel, but combusting fossil fuel is no longer a viable environmental option. There is nothing inherently beneficial to human civilization in using solar energy that was stored in fossil fuels some 300 million years ago—it was simply the best solution we had at the time.

Without predicting the results of scientific and engineering research before it is even funded, it is nevertheless difficult to imagine a sustainable future in which solar-thermal power stations do not play a key role in providing base load power.

This will not mean a lesser world.
It will just be a different world and it may even be a fairer world.

The Means

There are many scientists already following in the footsteps of Einstein, Tesla, Curie, Marconi and Edison. Today's exceptionally gifted individuals require the financial and infrastructure resources commensurate with the task of developing energy and fuel systems, as yet not even imagined.

Crucial also will be the economics of making it work. Those most talented in economics and finance will be needed to develop new economic paradigms that will enable rapid development and deployment of advanced zero emission technology.

We need to reform our use of the atmosphere as a 'no-charge' CO_2 emissions dump. Any discussion on this topic will inevitably raise political 'hot-button' issues such as taxation, regulation, rationing by price or coupon and perhaps even limiting some freedoms.

In the future, we may not have the freedom to do whatever takes our fancy, with no regard for the energy consumed and without paying an appropriate fee to compensate for damage to the biosphere.

Changing well-established business practices will not always be easy. It will involve changing some things that many of us currently see as vital to maintaining our comfortable lives. But, our way of life is going to change anyway. If we carefully manage this change, the future looks bright. Unpleasant change is a good deal more likely if we do nothing and let the biosphere decide how we shall be affected.

Roadblocks and Hurdles

There are hurdles to overcome before anything changes —we must firstly find the will. The lack of motivation for change is seen both in the upper echelons of power, where things can be made to happen and also throughout society.

The average person in the street does not want to see their daily routines or known expenses negatively impacted. However, this is where the drive for change may need to originate if the impetus for it does not come from the top.

Everyone, politicians, union officials, business and religious leaders and all of us need to take up the challenge and start tweeting.

There are vast amounts of money being made from greenhouse gas intensive industries and agricultural practices. Overcoming resistance to change from these quarters is a challenge we all face. It is also essential that one way or another, the media is persuaded to join the battle for the environment, or the fight will surely be lost.

Let's Not Become the New Dinosaurs

The frequent assertion of those denying anthropogenic climate changes is that levels of atmospheric carbon dioxide and temperatures have been higher in the past, so we therefore have nothing to worry about—this must be challenged at every opportunity.

There were periods in the distant past when carbon dioxide and temperatures were higher than today, but these were also the times when the environment was not favourable to the development of the advanced, highly organised and structured societies we live in and enjoy today. Sometimes these higher levels even pre-dated the rise of Homo sapiens.

This does prompt obvious questions:

- are we the new dinosaurs

- are we facilitating a climate model that will, in time, become so hostile that we don't have to wait for a wayward asteroid to seal our fate

- are we, in fact, busily organising our own demise

- is this fair to our children, grandchildren, great-grandchildren and ourselves.

This is a beautiful planet.
Let us ensure that present generations are not the last to enjoy it.

Start Now

At this moment in time, no one can be prescriptive about the actual solutions that will emerge to counter the developing crisis in the biosphere. None of us can anticipate exactly the result of scientific and engineering research that has yet to be fully funded.

In a complex system as large as a planet we must not wait for a major event, which is beyond any doubt linked to our changes to the chemistry of the atmosphere, before we bring in significant changes to the way we conduct ourselves. To do so will show a level of naivety beyond belief.

We must throw everything possible at the task of developing technologies to stop global warming and then the power of technological compounding will do the rest.

Remember—this is not a fight to save the planet, it may not be a fight to save the species, but it is definitely a fight to save our civilization and life as we know it. It is a fight not to be lost.

The time to start is now.

Author's Note and August 2018 Update

Change – Ready or Not is now an 'old' book; the origin of which dates back to the early 1990s. At that time, working in Australian manufacturing, I was able to observe first-hand some of the waste of physical resources that resulted from the absence of a CO_2 emission cost in the manufacturing cost base. This was evidenced by the premature scrapping of functional equipment with many years of still useful operating life — waste that could have been eliminated with a system for repurposing unwanted componentry in a different enterprise/application to efficiently utilise the already-paid-by-the-environment CO_2 emissions involved in their original manufacture. This concept has been expanded on in this book. It is an idea that should apply not only to industrial equipment, but also to consumer products and other CO_2 'emission hungry' items and services.

This book took a while to write — from 2006 to completion in 2015. In that time, consumer demands on the Earth's ecosystem have continued to increase significantly. The concepts put forward in this book are as relevant today as they were at the time of writing.

An individual can only detect the effects of global warming anecdotally. As individuals, we cannot sense a 1°C temperature difference or the effect on the planet of a 1°C global temperature rise over a sixty year period. We cannot 'feel' the effect of the change to the chemistry of the biosphere and the attendant warming of the planet on our skin. At any time, we experience only the local and current, temperature and weather.

The effect of the alteration to the chemistry of the biosphere is on a much larger scale — the climate zones of the entire planet. It is futile for human beings to attempt to compare the planet-wide changing climate with the immediacy of the day-to-day temperature and weather. The more reliable indicator is the longer term climate phenomenon — with the usual time span for quantifying climate trends measured by a thirty-year moving average.

Over the past thirty years, the daytime maximum temperature in Melbourne, Australia has ranged from 9°C and 46.4°C. Thus, changes in the day-to-day weather patterns give a thirty year temperature variation of 37.4°C from lowest to highest. Melbourne is typical of many cities around the world and all will have a maximum to minimum temperature variation. If we consider an even earlier time, circa 1950, the Earth's average temperature was 1°C less than today. For sake of discussion, if today is 23°C and under theoretically identical atmospheric conditions and wind patterns, it would have been only 22°C in 1950, we would we never measure that small difference on our skin. We could never remember how that temperature 'felt' 68 years ago as human beings are not calibrated with the accuracy and sensitivity of a scientific instrument for the measurement of temperature and pressure. Therefore, there is more opinion than hard data in the *human* assessment of how we should respond to the changing chemistry of Earth's atmosphere and oceans. This is why global data and rigorous scientific analysis is the only basis on which decisions about stabilising the biosphere can be made.

The rise in average global temperature is just one measure of the changing chemistry of the Earth's biosphere. Global warming and the changing climate, within and across climate zones, are the manifestation of those higher temperatures on a planet-wide scale. The alteration of Earth's climatic zones is far more complex — and dangerous. The thermodynamics of the atmosphere involves a study with many variables, but in general, as the temperature increases, the total energy within the system also increases. One result is there are weather events that have a greater local impact on some parts of the world than would otherwise be suggested by a mere 1°C temperature rise.

For instance, it has been well researched that increased ocean temperatures give rise to greater total energy within a cyclone (also called hurricanes or typhoons, depending on the geographic location). Unfortunately, when 'big-picture' climatic change clashes with and overwhelms the day-to-day local weather, the world experiences catastrophic weather events. This was the case in November 2013 when Typhoon Haiyan caused immense devastation in the Philippines and is why an analysis of the central barometric pressure of some recent cyclones would suggest that the rating scale should go beyond Category Six.

Thirty years ago, the late former British Prime Minister, Margaret Thatcher made a speech to the Royal Society in London, in which she eloquently outlined the threat that global warming posed to the environment in general and the climatic system in particular. Mrs Thatcher's speech is as relevant today as it was thirty years ago.

We cannot ignore the prescient warnings that were contained in this 1988 speech to the Royal Society and also in Mrs Thatcher's speech to the United Nations General Assembly in 1989. We certainly cannot ignore the chemical changes to the atmosphere for another thirty years. The melting of the Arctic Ice Cap, thawing of the Tundra's permafrost, reducing alkalinity in the oceans, more extremely hot days, fewer frosts and more so-called 'super-storms' combined, paradoxically, with less rainfall are not coincidences. Even small things are affected. As discussed in Chapter 7, research has identified a causal link between the warming trend and the emergence of a certain species of butterfly from the larvae ten days earlier than was the case for the same species in 1945.

A study of both large-scale and small-scale changes to our planet's climate zones and events arising from the changing chemistry of the planets biosphere makes for very sobering reading.

In real terms, little has been achieved to counteract global warming in the past thirty years. Weather events have become more extreme, even chaotic, and importantly the Earth's long established climate zones are changing. In particular, the warming of the northern regions of the planet is seeing ever more frequent breakdowns in the stability of the Polar Vortex. This is causing extremely high temperatures in some Arctic regions and extreme cold at lower latitudes as the Polar jet stream becomes more lobular and less circular.

Change – Ready or Not scopes possible solutions to our continuing alteration of the chemistry of the atmosphere and oceans and the enduring contest between our consumer society and the environment. This is a contest that will be heightened as increasing numbers of people rightfully demand to enjoy the benefits of consumer essentials and comforts presently enjoyed by many of us fortunate enough to live in a highly developed economy.

Commentary has shifted in recent years regarding the application of the terms climate change denier or sceptic; terms that are used in this book to describe those opposed to real economic action to drastically reduce CO_2 emissions. Those opposed to the concept that human activity is responsible for global warming/climate change have changed tactics. Their refrain is now more likely to be: "Of course I believe the Earth is warming and the climate is changing — it's just not caused by human activity". In a classic case of misdirection, the debate then becomes an argument about what is causing the warming.

The fact that atmospheric CO_2 — which makes up 99.5% of all dry atmosphere greenhouse gases — has increased over the Industrial Age by 45% is instantly dismissed as a possible cause by those who previously would have been called climate change deniers or sceptics.

We now hear it suggested that planetary and solar cycles are to blame or even, that global warming is caused by increased cosmic radiation arising from the Earth changing position in the Milky Way Galaxy — anything to deny that the 45% increase in atmospheric carbon dioxide is changing the chemistry of the atmosphere and causing the phenomenon we know as global warming or climate change.

There is no denying the myriad of cycles that affect planet Earth, but if we were to follow that line of reasoning, some of the most prominent and longer time-span temperature cycles suggest that without our human alteration to the chemistry of the biosphere, the Earth would more likely be entering a cooling rather than a warming cycle.

There will always be uncertainty in forecasting every possible eventuality in the atmosphere of a 5.98×10^{24} kg planet. Planetary climate physics cannot be that precise. But, we must always remain cognisant that any relatively small aberration at a planetary level can be catastrophic at a local level.

In terms of the continuing 'relative' stability of the biosphere, it is possible that we have been lucky so far. The role of the oceans, in providing a massive heat and carbon sink, may have benefitted us more than originally thought possible. This good fortune cannot continue indefinitely and the need for changes to the way we service and supply our burgeoning consumer demand continues.

We urgently need a realistic and appropriate globally applied cost for emitting CO_2 into the atmosphere as well as a globally applied total emission cost for every good and service. It is insufficient to simply put a sticker on a refrigerator or motor vehicle that states only the operating emission cost. What we urgently require is the total emission cost of getting that product or service out of the ground and into the consumer's possession.

When it is often cheaper and easier to go out and buy a brand new ink-jet printer rather than buy replacement ink cartridges, we have an illegitimate pricing policy which takes no account of the actual carbon cost of a new printer, compared to the relatively tiny cartridge. This small example should be considered in the context of the *repair and maintenance* strategies discussed in this book.

It is very likely that decentralised repair and maintenance, and all of the upstream/downstream ramifications of such a policy, could help to solve some of our society's current economic problems — in addition to the climate imperative.

Our present lack of sufficient action is not fair to ourselves, our children, our grandchildren and all the others who expect us to leave this planet in a better condition than at the time we inherited it.

Furthermore, it is insufficient to expect that citizen action alone can achieve the required change. Thirty years after it was first placed squarely on the global agenda it is now time for the leadership of governments around the world to step up to the task appropriately and sufficiently. If not now, then when?

A. G.

Glossary

Aerosol	Ultra-microscopic particles dispersed throughout a gas or a liquid.
Albedo	A measure of the proportion of the Sun's heat that is reflected back into space. Albedo varies with the particular features of the planet's surface.
Anthropogenic	In context of global warming, emissions of greenhouse gases that originate from human activities.
Argon (Ar)	An inert gas with a chemical symbol of Ar. Argon is the third most prolific atmospheric gas. Argon is not a greenhouse gas.
Base load power	The minimum power requirement that must be supplied throughout the entire energy demand cycle.
Biofuel	Energy from fuels that originate from living organisms, such as plants and animal fats.
Biosphere	The part of Earth occupied by living things, from the sediment in the deepest ocean to the highest spores, or other life, in the upper atmosphere.
Carbon dioxide (CO_2)	A colourless, odorless atmospheric gas with a chemical symbol of CO_2. CO_2 is the principle greenhouse gas keeping the Earth at a habitable temperature. To convert quantities of elemental carbon, in kilograms, to kilograms of carbon dioxide (CO_2) multiply by a factor of 3.67.
Carbon dioxide equivalent (CO_2e)	The combined warming potential of all atmospheric greenhouse gases in terms of the equivalent amount of carbon dioxide.
Cradle to grave or whole of life product cost	The total cost of each item, or service, that we consume, or is consumed on our behalf.
El Niño	Low or negative values of the Southern Oscillation Index (SOI).

235

Exponential growth/decay	Exponential growth, or decay, occurs when the rate of growth, or decay, of a mathematical function, commodity, organism, or population is proportional to its current value. This acts to compound the rate growth or decay.
Fossil fuel	Fuels formed by natural processes, from decaying organic material, over millions of years. They include coal, crude oil, methane and natural gas.
Geo-engineering	Cooling the planet by scientific and engineering intervention, usually above the Earth's surface.
Greenhouse Gases	An atmospheric gas that acts to prevent some of the Sun's heat being radiated back into space, in the same way as a traditional glass greenhouse; known as the 'greenhouse effect'.
Gross World Product (GWP)	The combined Gross National Product of every country in the world.
Habitat	The natural environment, or home, of any organism.
Kyoto Protocol	A 1997 International treaty that sets binding targets for greenhouse gas emissions for industrialised countries.
La Niña	High positive values of the Southern Oscillation Index (SOI).
Many-body problem	Microscopic particle systems containing a very large, or infinite, number of particles that interact within the system.
Methane (CH_4)	A greenhouse gas, both naturally occurring and of anthropogenic origin. Multiple sources of origin.
Nanotechnology	The technology to control individual atoms and molecules
Nitrogen (N)	A colourless and odourless gas comprising 78% of the Earth's atmosphere. Nitrogen is not a greenhouse gas.
Nitrous Oxide (N_2O)	A greenhouse gas, both naturally occurring and of anthropogenic origin, mainly from agricultural use of synthetic fertilizers.

Non-linear feedback	A system where the output from a process can influence the same process again in a future time, with a magnified effect.
Oxygen (O)	A colourless and odourless gas comprising nearly 21% of the Earth's atmosphere. Oxygen is not a greenhouse gas.
Particulates	Very small distinct solid particles that are suspended in a gas or liquid. In connection with global warming, the gas under consideration is usually the Earth's atmosphere.
Parts per million (ppm)	The number of parts of a particular constituent that is contained in a million parts of another. 1 part per million (1 ppm) = 0.0001 per cent (%).
Quantum electrodynamics	Quantum field theory describing interactions between photons and matter.
Quantum mechanics	The physics of atomic and sub-atomic phenomena.
Renewable energy	An energy source that does not make use of Earth's finite resources. For example solar energy from the Sun.
Solar thermal power	Power plant technology to use the sun's energy to generate electricity on a large commercial scale.
Southern Oscillation Index (SOI)	The SOI uses pressure differences in certain areas of the Pacific Ocean to indicate the likely formation and intensity of El Niño or La Niña events, which control rainfall in Eastern Australia.
Sulphur dioxide (SO_2)	A gas with a pungent odour released by volcanoes and various industrial processes and fuel burning, which act to counter the greenhouse effect.
Superconductor	A phenomenon of zero electrical resistance in a conductor.
Tipping Point	A condition that will cause some physical parameter to become unstable, often to the detriment of the circumstance being observed.

Explanatory Note

Throughout this book, the terms climate change and global warming are both used, but not completely interchangeably. Both are symptoms of rising CO_2 levels and global warming is a precursor to climate change. Similarly the terms 'atmosphere' and 'biosphere' are used, again not completely interchangeably. The biosphere includes the whole of the space occupied by living things from bottom of the sediment at the deepest point in the ocean to the highest that any spores or other life can survive in the atmosphere. The biosphere includes the majority of the atmosphere.

'Climate change denialist' and 'climate change sceptic' are used interchangeably as both are in common use in the community. Denialist is the term more frequently used and is generally favored over sceptic. A healthy scepticism is an integral part of the scientific method, but when used by 'climate change sceptics' it seems to become a denial of very compelling evidence. Both terms are used from the standpoint of common usage in the global warming/climate change debate rather than in the sense of a strict dictionary definition or interpretation.

Select Bibliography

Daly, Herman E., *Steady State Economics*: Island Press, Washington, D.C., 1991.

Francis, Jennifer, Department of Marine and Coastal Sciences, Rutgers University, New Jersey, United States of America.

Holmes, Edward (Ted): Management Advisor, Academic, Poet, 2009.

Kearney, Michael, Briscoe, Natalie J., Karoly, David, et al., *Early Emergence in a butterfly causally linked to anthropogenic warming*: Biology Letters, Journal of the Royal Society, March 2010.

Kennedy, John F., *Special message to Congress on urgent national needs*: Presidential Papers, President's Office Files, 25th May 1961.

Thatcher, Margaret, *www.margaretthatcher.org/document/107346*: Margaret Thatcher Foundation, www.margaretthatcher.org.

Ummenhofer, Caroline, England, Matthew, et al., *Geophysical Review Letters*: University of New South Wales, Climate Change Research Centre, 2009.

Select Bibliography

www.ingramcontent.com/pod-product-compliance
Lightning Source LLC
Chambersburg PA
CBHW060358220326
41598CB00023B/2962